Capturing the German Eye

Capturing the German Eye

American Visual Propaganda in Occupied Germany

Cora Sol Goldstein The University of Chicago Press
Chicago and London

Cora Sol Goldstein is associate professor of political science at California State University, Long Beach.

The University of Chicago Press, Chicago 60637
The University of Chicago Press, Ltd., London
© 2009 by The University of Chicago
All rights reserved. Published 2009
Printed in the United States of America

18 17 16 15 14 13 12 11 10 09   1 2 3 4 5

ISBN-13: 978-0-226-30169-3 (cloth)
ISBN-10: 0-226-30169-9 (cloth)

Library of Congress Cataloging-in-Publication Data
Goldstein, Cora Sol.
   Capturing the German eye : American visual propaganda in occupied Germany / Cora Sol Goldstein.
     p. cm.
   Includes bibliographical references and index.
   ISBN-13: 978-0-226-30169-3 (cloth : alk. paper)
   ISBN-10: 0-226-30169-9 (cloth : alk. paper)
   1. Germany—History—1945–1955. 2. Propaganda, American—Germany. 3. Psychological warfare—Germany. I. Title.
DD257.G596   2009
327.1'4094309044—dc22

                                             2008031299

∞ The paper used in this publication meets the minimum requirements of the American National Standard for Information Sciences—Permanence of Paper for Printed Library Materials, ANSI Z39.48-1992.

This book is dedicated to my father, Daniel J. Goldstein

Contents

Acknowledgments ix
Abbreviations xiii

Introduction 1
One American Atrocity Propaganda 21
Two American Propaganda Films 41
Three ICD's Blind Spot: The Fine Arts 69
Four Overt and Covert American Actions in the German Fine Arts 89
Five Iconoclasm and Censorship 105
Conclusion 127

Notes 135
Bibliography 173
Index 193

Illustrations follow page 82

Acknowledgments

In the late 1990s, when I began the research that led to this book, American military occupations appeared to be a relic of the past, a domain of historical studies that could scarcely be expected to provide new insights or to have contemporary policy implications. Moreover, I approached the issue of military occupation in an unorthodox manner, by looking at the control of visual culture. Fortunately, the University of Chicago offered an interdisciplinary community that encouraged me to cross intellectual boundaries and allowed me to interact with scholars in a variety of disciplines—political science, history, German studies, cultural studies, and art history. Still, without the enthusiastic backing of Michael Geyer, Sander Gilman, William H. Sewell Jr., and Lisa Wedeen this project would never have survived.

I am also grateful to Michael Dawson, Gary Herrigel, John Mearsheimer, Ronald G. Suny, and Stephen Walt, of the Department of Political Science, who helped me navigate graduate school and always provided help when needed. I am indebted to Kimberley Rorshack and Carroll Joynes for their support. I profited greatly from the University of Chicago Workshops, in particular the Comparative Politics Workshop, the Modern European History Workshop, and the Cultural Policy Workshop. I remember my years at Wilder House with great fondness, and in particular enjoyed sharing space and friendship with Elise Giuliano, Devin Pendas, and Alex Thompson.

In a sense, the origin of this project can be traced to the University of California at Berkeley, where I took my first German history and German politics courses and became fascinated with Germany. Erhard Stölting, then a visiting professor at Berkeley, guided me through the intricacies of post-Nazi German politics. And during my earliest fieldwork in Germany, Erhard and Angela Stölting's home in Berlin became my home away from home.

I am indebted to Volker R. Berghahn, David Caute, Jeffrey Herf, Harold Hurwitz, Christof Mauch, and Robert Vitalis, for their confidence in me and in this project.

I am also grateful to my colleagues in the Department of Political Science at California State University, Long Beach. Charles Noble, the chair of the department, became a mentor and friend, and strongly encouraged my research agenda.

Lilo Grahn-Sandberg, the Grossman family, Joseph Koerner, Carl and Christopher Lehmann-Haupt, Leonard Linton, Juliana Munsing, Alfred H. Paddock Jr., Sandra Schulberg, and Hans N. Tuch shared with me information, photographs, and personal documents.

I also thank the archivists and librarians at the University of Chicago, the Art Institute of Chicago, the Museum of Modern Art, the National Gallery of Art, the Archives of American Art, the U.S. Army Center of Military History, the Truman Library, the University of North Carolina at Chapel Hill, the Library of Congress, the U.S. Holocaust Memorial Museum, the U.S. National Archives, the State Department, Vassar College, the University of Texas at Austin, the Academy of Motion Picture Arts and Sciences, the Landesarchiv Berlin, the Kunstbibliothek, the Staatliche Museen zu Berlin, Chronos, the Deutsches Historisches Museum, the Gedenkstäte Buchenwald, the Zentralinstitut für Kunstgeschichte Munich, the Neue Nationalgalerie Berlin, the Archiv für Bildende Kunst Nürnberg, the Kinemathek Hamburg, the Altes Rathaus Steglitz, the Haus am Waldsee, the Mahn-und Gedenkstätten Wöbbelin, the KZ-Gedenkstätte Dachau, the Deutsche Film-und Fernsehakademie Berlin Bibliothek, the Landesbildstelle Berlin, the Staatsbibliothek zu Berlin, the Imperial War Museum, and the Centre Pompidou.

The project was funded at different stages by grants from the Deutscher Akademischer Austauschdienst, the MacArthur Foundation, the University of Chicago, the German Historical Institute (Washington, D.C.), the Gottlieb Daimler und Karl Benz Stiftung, the Center for Arts and Culture (Washington, D.C.), the Horowitz Foundation for Social Policy, the American Political Science Association, and California State University at Long Beach. Earlier versions of chapters 3, 4, and 5 were published in *Intelligence and National Security, Diplomatic History,* and *German Politics and Society.*

I want to express my profound thanks to the readers at the University of Chicago Press, Rebecca Boehling and Liliane Weissberg, for their careful consideration of successive drafts of this book. Their insightful comments, detailed criticisms, and thoughtful suggestions helped me enormously. Senior editor Susan Bielstein, her editorial associate, Anthony Burton, and project editor Joel Score patiently helped out a rookie author. Their assistance has been invaluable.

It is to my family, however, that I am most indebted. My mother, Cora Sadosky, and my father, Daniel J. Goldstein, have shaped my intellectual trajectory as much as they have influenced who I am as a person. I am grateful to them for the entire winding path, which began in Buenos Aires. I am also grateful to my husband, Tom Johnson, and to our daughter, Sasha Malena Johnson. Tom and I met before this project began, when I was carrying my B.A. thesis everywhere I went. He thought that was actually fantastic, and he continues to read, critique, and encourage my work. Our daughter was born around the time that I sent this book to the University of Chicago Press for consideration. She is beautiful, sweet, and full of life, my ray of sunshine.

Abbreviations

CAD	Civil Affairs Division, U.S. War Department
CDU	Christlich Demokratische Union Deutschlands (Christian Democratic Union of Germany)
CIA	Central Intelligence Agency
CSU	Christlich-Soziale Union, Bayern (Christian Social Union, Bavaria)
DFU	Documentary Film Unit, ICD
E&CR	Education and Cultural Relations Division, OMGUS
ETOUSA	European Theaters of Operation, U.S. Army
HICOG	High Commissioner for Germany
HUAC	House Un-American Activities Committee
ICD	Information Control Division, OMGUS
ISB	Information Services Branch, OMGUS
JCS 1067	Joint Chiefs of Staff Directive 1067
JCS 1779	Joint Chiefs of Staff Directive 1779
KPD	Kommunistische Partei Deutschlands (German Communist Party)
MFA&A	Monuments, Fine Arts, and Archives Section, OMGUS
MID	Military Intelligence Division
MoMA	Museum of Modern Art, New York
NARA	U.S. National Archives
OIC	Office of International Information and Cultural Affairs, U.S. Department of State
OMGUS	Office of Military Government U.S. in Germany
OSS	Office of Strategic Services
OWI	Office of War Information
PWD	Psychological Warfare Division
SHAEF	Supreme Headquarters of the Allied Expeditionary Force
PWD/SHAEF	Psychological Warfare Division of the Supreme Headquarters of the Allied Expeditionary Force
SED	Sozialistische Einheitspartei Deutschlands (Socialist Unity Party of Germany)
SMAD	Sowjetische Militäradministration in Deutschland (Soviet Military Administration, Germany)

SPD	Sozialdemokratische Partei Deutschlands (Social Democratic Party of Germany)
USFET	U.S. Forces, European Theater
USGCC	U.S. Group Control Council, Germany

Introduction

May 8, 1945—V-E Day—marked the end of World War II in the European Theater of Operations, but not the beginning of peace. By the end of 1946, the confrontation between the capitalist world and the Soviet Union had resumed following a wartime hiatus. Over the next four decades, the United States and the Union of Soviet Socialist Republics would avoid direct military confrontation but clash repeatedly in proxy wars fought by client regimes and rival guerrilla movements. In the absence of overt war, the two superpowers designed and carried out campaigns to influence the opinions, emotions, attitudes, and behavior of target populations, and to limit the reach of the other power in key regions of the world. Psychological warfare, also known as strategic propaganda, was a key factor in the Cold War.[1]

Occupied Germany was the first battlefront in this postwar conflict. In 1945 the four Allies—the United States, the Soviet Union, the United Kingdom, and France—established control over the German media in their respective zones of Germany and sectors of Berlin. Yet despite this quadripartite rule, the main contenders in the political struggle for ideological and cultural hegemony in occupied Germany were the Americans and the Soviets. By 1949 the Office of Military Government U.S. in Germany (OMGUS) and the Sowjetische Militäradministration in Deutschland (SMAD) were using all avenues of mass communications and cultural affairs—newspapers, journals, fiction and documentary films, posters, radio, music, literature, theater, caricature, and the fine arts—to disseminate their competing messages.

Neither the classic studies of the American occupation of Germany nor the recent political science literature on democratization by force have analyzed the role of cultural policy as strategic propaganda. Harold Zink and John Gimbel, in their seminal studies of OMGUS, focus on the ways in which the American military government in Germany changed economic, financial, and political institutions in the immediate postwar period.[2] Yet they never consider the issue of cultural policy as a component of psychological warfare. In fact, Zink, the official historian of OMGUS, claims that the American military government's film policy was not a means of propaganda.[3] Recent studies on democratization by force, which often posit Germany as a model of success, also bypass the issue of strategic propaganda.[4] The assumption seems to be that the key to imposing democracy is changing the institutional framework.

It is only recently that scholars interested in the history of Germany in the immediate aftermath of World War II have begun to unravel the role of information control and propaganda in the OMGUS project. There are now thorough studies on American atrocity propaganda, press policy, film policy, music policy, art policy, and theater policy in occupied Germany.[5] Moreover, debates on the "Americanization" of Europe and the role of the Central Intelligence Agency's cultural policies during the Cold War have revived interest in the ideological battle between the United States and the USSR. Indeed, the end of the Cold War has helped clear the way for such research. To compare American propaganda policies in postwar Germany with those of the Nazis or the Soviets would, during the Cold War, have been deemed preposterous or anti-American. Now, however, it is evident that the United States also used culture as a political weapon.[6]

In this book I trace the development of American visual propaganda in occupied Germany from 1945 to 1949. During this period the U.S. Army and OMGUS produced successive visual narratives of German fascism and of German-American relations, employing images to attack and discredit first the Nazis and later the Communists. The content changed as American strategic and tactical needs in Europe shifted, but visual propaganda remained essential to American denazification and reeducation efforts. The Americans censored images and relied on film, photography, and the fine arts to project their political agenda and to seduce, dissuade, and control the German population in the American zone and sector.

My goal is to highlight the importance of information control and mass propaganda in radical regime change and to illustrate the role of the visual in mass propaganda. In the television era, it is inconceivable to think of psychological warfare without images. But visual propaganda has always been as important as verbal messages. Images are often more effective than words in capturing the attention of the viewer, and they are better able than written and oral texts to crystallize sentiments. Images simplify and convince. If something is visually represented, viewers tend to assume that it happened, that it exists, that it is true. Visual images are overwhelmingly important in the interpretation of our conscious world, shaping the way we consider ourselves and others. The suppression of familiar visual elements attenuates memories; the continuous iteration of images creates new memories. Through either mechanism, visual propaganda can subtly, yet rapidly, change our points of reference and lead to profound changes of worldview.[7]

It is well known that the Third Reich placed great effort into mass indoctrination through visual propaganda. The Nazis banned films and art that did not conform with their racist ideology. In 1936 they purged German museums and galleries of modern art, and in July 1937 organized the "Degenerate Art Exhibit" to show the German people that modern art was the product of the infirm, the insane, and the decrepit. At the same time, the regime flooded the visual domain with images that implicitly or explicitly promoted Nazi ideals—mass spectacles, portraits and statues of the Führer, Nazi paraphernalia, and monumental buildings, as well as films, photographs, posters, and art. This radical remodeling of the visual sphere set limits to the imagination; anything that differed from the normative paradigms of the regime was deemed abnormal. By controlling what the German people saw and the context in which they saw it, the National Socialist government attempted to create a coherent new world devoid of dissonance and heterogeneity.

During World War II, the U.S. government had also relied heavily on visual propaganda. In June 1942 the Roosevelt administration created the Office of War Information (OWI) to shape American public opinion and encourage prowar sentiments. The agency sought to explain the war to American civilians, to mobilize the home front, to increase morale, and to indoctrinate American soldiers. Elmer Davis, the director of the OWI, believed that the easiest way to transmit propaganda effectively was through film, especially when spectators were unaware that they were being propagandized. Therefore, Hollywood was conscripted for the war effort. Under the guidance of the OWI, Hollywood produced feature films intended to educate, inspire, and entertain, and military training films to teach soldiers why they were fighting and how to fight. The relationship between the OWI and Hollywood amounted to a virtual merger between the state and the movie industry with the goal of shaping American public opinion.[8]

In April 1945, in the immediate aftermath of the liberation of the Nazi extermination and concentration camps, American visual propaganda had its greatest impact both at home and in Germany. Photography made the concentration camps real to the American public and constructed a new image of German criminality. In Germany, the U.S. Army designed the "atrocity propaganda policy," an exercise in politics and punishment through visual means. Through photography, film, and forced visits to the camps, Germans were made to *see* Nazi crimes. A central goal was to replace the image of the Third Reich constructed by the Nazi propaganda machine with the image of mass atrocity. Months later, the Americans used visual material gathered at the camps—films, photographs, and objects—as evidence in the Nuremberg trials. The photographic image of the concentration and extermination camp system became, retroactively, the defining representation of the Nazi genocidal project.

Once Germany was occupied, the U.S. Army, and then OMGUS, conducted a massive campaign of iconoclasm targeted at eliminating the legacy of the Nazi regime from the visual realm. Visual remnants of the Nazi era—paintings, monuments, statues, emblems, and military symbols—were eliminated from public view. More than eight thousand works of art were removed from museums and government buildings, classified as Nazi or militaristic, and sent to the United States to be stored in military depositories. The American military government sequestered all films produced by Nazi Germany and classified them as poisonous material. This was a campaign of radical visual censorship.

Moreover, the U.S. military government used the American experience in visual indoctrination and propaganda to sell private enterprise, consumerism, democracy, and liberal capitalism in occupied Germany. The Americans were the global leaders in film, advertising, and marketing in the 1940s, and film was central to this process of indoctrination. In July 1945, OMGUS, following SMAD's lead, began to reopen movie theaters in the American zone and sector. The OMGUS Information Control Division, the State Department, and the War Department selected the films that would be shown in the American zone and sector. These American films were meant to advance the U.S. goals in Germany by challenging the Nazi stereotypes about America, displaying a positive image of American society, and highlighting the structure of American democracy. Entertainment was a secondary concern; the priority was to introduce new political conceptions and normative values to the occupied population. By 1947, OMGUS begun using film as part of its anticommunist propaganda campaign.

While the Americans were quick to use photography and film as visual propaganda, they took much longer to develop an art policy. The fine arts were outside the parameters of conventional visual propaganda. Yet in the German context, the fine arts were already politicized. The Nazis had eliminated modern art from museums and used their own art as political and ideological propaganda. The Soviets, for their part, entered Germany with a blueprint for cultural warfare already in place. By 1946 the Soviets had focused their cultural propaganda efforts on three main themes: the social responsibility of artists and intellectuals, the denigration of modern art, and the glorification of Socialist Realism in the arts. OMGUS was slow and tentative in its reactions to the SMAD cultural offensive. The War Department, which directly controlled OMGUS, was indifferent to the concept of international cultural warfare and failed to appreciate the importance of cultural politics in the postwar German context. Nonetheless, by 1946 a small group of cultural officers in OMGUS had begun to establish bonds with German modern artists to further German-American relations and stimulate the reintroduction of modern art. They relied on personal contacts, informal networks, and covert funding of cultural organizations. The American art policy in Germany, constrained and often vacillating as it was, is politically important because it prefigured the cultural policies of the CIA during the rest of the Cold War.

The American visual propaganda policy in occupied Germany was shaped by two factors: the fear of alienating the German population and the constraints imposed by American domestic politics. In seeking to make the American message credible, but to avoid antagonizing the Germans, OMGUS gradually reduced its atrocity propaganda campaign. OMGUS policies were also affected by the end of the fragile and unstable wartime alliance between the American right and the American left. In 1945 a newly elected Republican-dominated Congress pushed American domestic politics to the right. Using anticommunism as a pretext, the Republican majority, with support from southern Democrats, began the long fight to undo the New Deal, eliminate New Dealers and leftists from the administration, and rewrite the social contract in America. Because OMGUS was under intense scrutiny from the U.S. Congress, this shift in American politics conditioned its response to Soviet anti-American offensives on two key propaganda issues—race and modernism in art. The American project in Germany, although ambitious and ultimately successful, was shaped and constrained not only by the emerging Cold War, but also by domestic cultural battles over the definition of an exportable image of America.

The evolution of American intelligence

World War II and the occupation of Germany led to the transformation of the American intelligence apparatus from an exclusively military outfit with strong anti-intellectual leanings to a predominantly civilian operation able to engage in cultural warfare worldwide. This evolution culminated in the creation of the Central Intelligence Agency in 1947. The CIA was endowed with the triple responsibility of gathering and analyzing foreign intelligence, carrying out covert offensive operations, and designing and projecting psychological warfare campaigns abroad.

Military Intelligence Division

During the first half of the twentieth century, American military intelligence was led by the fanatically anticommunist, anti-Semitic, and xenophobic Military Intelligence Division (MID).[9] Many intelligence officers were nativists who believed in the superiority of the "Nordic race" and regarded as true Americans only those with Anglo-Saxon ancestry. In 1919, after the Russian Revolution, the leaders of MID became obsessed with the idea that Communists, Jews, and Jewish international financiers—the European illuminati—were bent on terror and bloodshed and colluding to take over the world.[10] Since many liberal and left-wing American intellectuals were enthusiastic admirers of the Russian Revolution, MID responded to the events in the USSR by launching a massive anticommunist, anti-Jewish, and anti-intellectual campaign at home, developing a domestic espionage network to identify "internal enemies" and free the United States of "subversive" influences.

In the 1920s Army intelligence continued to investigate the connection between American Jews, the international Jewish movement, Jewish finance, and Bolshevism. Simultaneously, MID lobbied the U.S. Congress to limit the growing ethnic diversity of the nation, which it considered to be a strategic danger. MID officers established alliances with anticommunists in the State Department, with J. Edgar Hoover and the FBI (then called the Bureau of Investigation), with conservative members in the U.S. Congress, and with the leaders of the American eugenics and anti-immigration movements. MID also launched a disinformation campaign to disseminate anti-Semitic and anticommunist propaganda. The rise of the Third Reich in the 1930s did not cause MID to modify its agenda. Its experts on Germany concluded that the Nazis were preparing to wage war against the East, not the West, and saw no threat to the United States.[11]

Given the characteristics of MID, it is not surprising that President Roosevelt created a parallel intelligence service as soon as the United States entered World War II. The two main American intelligence organizations during the war were the Office of Strategic Services (OSS) and the Psychological Warfare Division, Supreme Headquarters, Allied Expeditionary Force (PWD/SHAEF). The OSS operated at the strategic level, PWD/SHAEF at the theater level. OSS and PWD/SHAEF, unlike MID, employed both military and civilian personnel. The civilian contingent formed a highly heterogeneous group in terms of national origin, social class, and expertise.

Office of Strategic Services
Created in June 1942, the Office of Strategic Services revolutionized American intelligence. The OSS was charged with three tasks: the collection and analysis of foreign political, psychological, sociological, and economic data; the development of an American psychological warfare program against the Axis powers; and the design and implementation of covert operations in Nazi-occupied Europe. To head the new agency, Roosevelt chose his personal advisor on intelligence, William "Wild Bill" Donovan, a World War I veteran and successful Wall Street lawyer.[12]

The mission of the OSS was to combat fascism, not Communism. Donovan's OSS became, to the horror of Army intelligence, an eclectic employer that recruited heavily among sectors of the population that the MID considered anathema: intellectuals, Jews, New Dealers, leftists (including communists), and foreigners (often European expatriates who were Jewish and left-wing). The OSS was formally subordinate to the Joint Chiefs of Staff, but its relations with the U.S. Army were at best strained. Major General George V. Strong, who directed the MID during World War II, was a staunch opponent of the OSS, claiming that Donovan's organization conflicted with the interests of the Army.[13]

The OSS was divided in branches—Morale Operations, Special Operations, Foreign Nationalities, Secret Intelligence, and Research and Analysis. Morale Operations disseminated "black" propaganda—information (which could be either true or false) from unacknowledged sources. Special Operations carried out missions in enemy-occupied territory, ranging from sabotage to morale operation plans. Foreign Nationalities used anti-Axis immigrant groups based in the United States to wage political warfare abroad. Secret Intelligence, modeled after the British Secret Intelligence Service, was in charge of spying and counterintelligence. Research and Analysis, headed by William Langer, was mostly staffed by university professors and lawyers working on research projects. Langer had a staff of approximately nine hundred scholars, including prominent historians, economists, political scientists, geographers, psychologists, and anthropologists. Ruth Benedict, Howard Becker, Margaret Mead, Alex Inkeles, Walter Langer, Douglas Cater, Herbert Marcuse, Arthur Schlesinger Jr., Walt W. Rostow, Edward Shils, H. Stuart Hughes, Gordon Craig, Crane Brinton, John King Fairbank, Sherman Kent, and Ralph Bunche all worked for the OSS. Many of the OSS experts on German questions were German emigrés who had become professors at the New School for Social Research in New York. These people spoke fluent German and had firsthand experience with German culture and politics. They also embodied everything that MID most distrusted and feared.

In 1944, at the peak of its power, the OSS employed nearly thirteen thousand men and women.[14] On October 1, 1945, President Truman disbanded the OSS, and many of its functions were later taken over by the Central Intelligence Agency. The CIA followed the OSS model closely—it too recruited intellectuals and academics.

Psychological Warfare Division

On June 25, 1941, John J. McCloy, then assistant secretary of war, created the highly secret Psychological Branch within the War Department General Staff G-2, headed by Lieutenant Percy Black. McCloy was interested in the Nazi concept of *Weltanschauungskrieg,* the use of psychological actions (including terror and propaganda) to control populations, both domestic and foreign. McCloy believed that the American experience in peacetime propaganda—political and commercial—could be used in wartime to manipulate the domestic front and to weaken the resolve of the enemy. The Psychological Branch was later renamed the Special Study Group, and in March 1942 it become the Psychological Warfare Branch, G-2.[15] In December 1942 the branch was dissolved and the Special Study Group/Psychological Warfare Branch was formed. For a short period it was incorporated into the OSS, after which it returned to the War Department.

During the war, the control and execution of psychological warfare was displaced from Washington, D.C., to the theaters of operation. Each American expeditionary force was endowed with its own Psychological Warfare Branch (PWB). In 1942 the Allied Forces Headquarters in North Africa had the largest of these units (PWB/AFHQ). When General Dwight D. Eisenhower began the preparations for D-Day, PWB, highly expanded and in close contact with the OSS, became the Psychological Warfare Division, Supreme Headquarters, Allied Expeditionary Force (PWD/SHAEF).[16] PWD/SHAEF, led by Brigadier General Robert A. McClure, had as its mission to disseminate propaganda in the European Theater of Operations, to demoralize the enemy and break their will to fight, and to sustain the morale of American troops and their supporters in countries occupied by the Nazis.[17]

PWD/SHAEF, much like the OSS, was a highly diverse organization with both military and civilian personnel. The division included personnel from the Office of War Information and the OSS, and from the British

Political Warfare Executive. General McClure reported directly to General Eisenhower, and he appointed military officers to direct the Intelligence and Liaison sections of the organization. McClure recruited among the graduates of the Psychological Warfare School at Camp Ritchie, Maryland, because these soldiers had training in psychological warfare and were fluent in foreign languages. He also employed civilians with a penchant for intelligence, even if they lacked formal training in psychological warfare. The American military officers in PWD/SHAEF saw the civilians "sykewarriors" of PWD and the OSS as a group of eccentric intellectuals.[18]

During Operation Overlord, the Normandy invasion in 1944, PWD/SHAEF was responsible for psychological warfare operations in the combat zone, and for strategic propaganda in Nazi-occupied Europe and Germany. The division employed a mixed menu of covert, overt, and "black" propaganda carried out through massive leaflet campaigns and the radio. In the final phase of combat, PWD/SHAEF prepared the ground for Germany's unconditional surrender through obedience training. The objective of this effort was to persuade German civilians and soldiers to obey the orders issued by the Allied occupation and abandon any illusion of resisting the invading forces.[19] Lastly, PWD/SHAEF was ordered, months before Germany's surrender, to create an apparatus of information control for occupied Germany.

Once American troops reached German territory, PWD/SHAEF had the responsibility of dealing with all the political and propaganda tasks required by the American occupation. As soon as Germany was occupied, the sykewarriors became involved in information control. They disassembled the remains of the Nazi propaganda apparatus, and began their own propaganda agenda. By the end of the war, PWD/SHAEF had a staff of 460 men and women.[20] After the unconditional surrender of the Third Reich, the OSS was dissolved and much of the staff of PWD/SHAEF was incorporated into the Information Control Division (ICD) in Germany. The new outfit, which included former OSS officers, took charge of psychological warfare in the American zone and sector after the occupation of Germany.

The American occupation of Germany

Defeat sunk Germany into a humanitarian quagmire. Practically all of its main cities had been severely damaged by the Allied aerial bombardment, and large sectors of Berlin had been reduced to rubble by ground fighting.[21] The collapse of the Nazi Party left the country without civil administration. The economy was crippled, there were few consumer goods, the German currency had ceased to exist, and cigarettes became the true unit of monetary exchange. Public transportation was in ruins, as were the communication networks. Five million Germans who had inhabited the former eastern territories had to be accommodated in an area that was one-third smaller than pre-1938 Germany. Millions of people were homeless or lived in ruined houses, facing devastating shortages of coal, gas, and electricity. The food supply was minimal, the population was famished, and the winter of 1945–1946 was particularly harsh.

After the defeat of the Third Reich, Germany was divided into four zones of occupation. The United States controlled southern and west-central Germany (Hesse, Bavaria, and part of Baden-Württemberg); the Soviet Union, the eastern area (Saxony, Thuringia, and Mecklenburg); Great Britain, the northwestern part (Lower Saxony, North Rhine-Westphalia, and Schleswig-Holstein); and France, the southwest (part of Baden-Württemberg and the Saarland west of the Rhine). There were approximately twenty-six million Germans in the American zone, a similar number in the Soviet zone, twenty-two million in the British zone, and close to six million in the French zone. The largest cities in the U.S. zone were Frankfurt and Munich. Berlin was divided into sectors. The Soviet sector had the largest population with over a million people; the American sector had just under a million, the British sector six hundred thousand, and the French sector four hundred thousand. An Allied Control Council had authority over the country as a whole, but each supreme military commander had executive power in his zone.

OMGUS was established five months after Germany's unconditional surrender, on October 1, 1945.[22] General Lucius D. Clay was responsible for American policy in occupied Germany, first as deputy military governor and then as military governor. Although the U.S. Army had had some experience in military government in Cuba and the Philippines, no military government in the history of the United States had confronted the political complexities that OMGUS managed on a daily basis between 1945 and 1949. General Clay had a dual mission: to guarantee stability in a chaotic social situation, while simultaneously implementing the "Four D's"—denazification, democratization, demilitarization, and decentralization. The first of these immense tasks entailed guaranteeing security, suppressing military insurgency, preventing civil insurrection, feeding the population, avoiding sanitary disasters, and reactivating the economy in the American zone and sector. The second, implementing the Four D's, was tantamount to orchestrating a political, social, and cultural revolution.

Yet domestic disagreements over what to do with Germany made planning difficult and contentious. The War Department, under Secretary Henry Stimson, and the State Department, under Secretary Cordell Hull, favored a "soft peace." They gave priority to reconstruction of the German economy, suggested that Germany pay moderate reparations, and lobbied for German political unity. In contrast, Henry Morgenthau Jr., the secretary of the treasury, championed a "hard peace." Morgenthau considered denazification a generation-long project and favored destroying German heavy industry, arguing that "Germany's road to peace leads to the farm." The tension between the two approaches was not fully reconciled by the time of the occupation.[23]

The central goal of democratization created a paradoxical situation. A military government involved in nation-building is, by definition, an authoritarian regime. Its success in imposing new values depends, in large part, on maintaining a monopoly of violence. Force is needed to guarantee law and order, and to make possible the control and manipulation of information in the occupied area. Carl J. Friedrich, the German-born director of the Harvard School of Overseas Administration (which trained American military personnel to participate in military governments) and later General Clay's constitutional and governmental affairs advisor (1947 to 1948), tried to resolve the contradiction inherent in the concept of democratization by force. He argued that OMGUS was a "constitutional dictator aiding in the reestablishment of constitutional democracy rather than dictating democracy."[24] According to Friedrich, a military government run by a constitutional democracy, unlike a conventional dictatorship, progressively relaxes repression and moves toward the establishment of a constitutional system. Friedrich acknowledged that OMGUS censored and repressed but claimed that it did so to constrain antidemocratic elements.

In 1945 the American military government allowed the German population very limited freedom and exerted an unprecedented degree of political control. Joint Chiefs of Staff Directive 1067 (JCS 1067), the American military directive that informed OMGUS policy from 1945 to 1947, explicitly rejected the idea that the United States was liberating a population held captive by a dictatorship. Instead, it stated, Germany "will not be occupied for the purpose of liberation but as a defeated enemy nation." Germans were to be controlled and monitored:

a. No political activities of any kind shall be countenanced unless authorized by you. . . . b. You will prohibit the propagation in any form of Nazi, militaristic, or pan-German doctrine. c. No German parades, military or political, civilian or sport, shall be permitted.[25]

Freedom of religious worship and freedom of speech were allowed only to the extent that they did not jeopardize America's military and political priorities.

Within these limits, the Americans did encourage German political life. OMGUS established competitive political parties and an independent judiciary and fostered grassroots political organizations. In early 1946 General Clay instructed American political officers to organize elections in *Landkreise* (rural counties), *Stadtkreise* (urban counties), and *Landtage* (state legislatures). Time was of the essence. Clay wanted to restore German self-government as soon as possible because, he believed, the mood at home would make a long and expensive military occupation impossible.[26]

American control of information, 1945–1946

After V-E Day, PWD/SHAEF became the Information Control Division (ICD) in Germany. General McClure became the head of the new outfit and kept much of the PWD/SHAEF personnel. At first ICD was independent from the military government, but in February 1946 it became fully incorporated into OMGUS and McClure began to report directly to General Clay. The main objectives of ICD, central to the American denazification and democratization efforts, were to design and project consolidation propaganda in occupied Germany, to make Germans cooperate with the U.S. Army and the American military government, and to generate a current of public opinion favorable to the American agenda. On the one hand, ICD had to suppress Nazi and militaristic influences on German culture and in the German media. On the other, it tried to introduce the principles of liberal democracy that the Americans associated with their political system.[27]

To advance these goals, ICD carried out two simultaneous policies—one negative and one positive. The negative policy was based on censorship and suppression. The division monitored all aspects of cultural production—specific sections controlled newspapers, journals, film, literature, theater, and music—and prohibited fascist, militaristic, and nationalistic messages that could inflame pro-Nazi sympathies and encourage resistance to the American project. Although there was no specific fine arts section, the fine arts were supervised by the Theater and Music Section. ICD personnel screened and vetted authors, playwrights, journalists, artists, museum curators, art dealers, producers, actors, and directors, and excluded those deemed politically tainted by their Nazi past. In so doing, ICD defined and policed the boundaries of the acceptable and the desirable in the American zone and sector.

The positive ICD policy initially focused on the press and film. During the first year of occupation, ICD licensed a politically and ideologically heterogeneous group of German newspaper and journal editors interested in the creation of a new democratic German press. At the same time, OMGUS reintroduced American films, albeit in a highly selective fashion. Films that portrayed the United States in anything but a good light were not shown.

While ICD was focused on the reconstruction of a democratic German press and used American film as mass propaganda, a broader German cultural revival was not an immediate priority for OMGUS. JCS 1067 did not mention cultural revival, and the higher echelons of OMGUS did not consider "high culture" a political weapon. Neither General Eisenhower nor General Clay were particularly interested in cultural affairs, nor were they personally involved in German cultural life. The high officers of the American military government—drawn from the army, industry, and the business world—had not been selected for their familiarity with German history and culture. Some excelled in macroeconomic management and others had strong ties with German finance and industry, but they had not been trained as political cadres and had no background in cultural policy.

However, OMGUS also had civilian officers, many of them in ICD, who were intimately familiar with German culture. Some were German exiles and expatriates, and others were American scholars versed in German *Kultur*. Indeed, during the first two years of the occupation, OMGUS was a complicated and diverse organization. Career military officers, citizen-soldiers who had enlisted to defend democracy and destroy fascism, and civilian employees seeking a role in the reconstruction of Europe, all coexisted within OMGUS. As the case of the German fine arts clearly shows, it was often small groups of OMGUS cultural officers, working in the periphery of the organization, who organized overt and covert operations in the cultural field.

Politics and culture in Soviet-occupied Germany

In 1945 the Soviets, like the Americans, talked about bringing democracy to Germany. The Soviet military government and German Communists returning from exile in Moscow, professed a commitment to establishing the foundations of an antifascist and democratic Germany. The Communists interpreted Nazism as a product of advanced capitalism, and claimed that if postwar Germany embraced the Western capitalist model of society, it would return to fascism. According to Soviet propaganda, the emergence of a new German culture—anti-Nazi, anticapitalist, and socialist—was the key to the moral, ideological, and political regeneration of Germany. For the Communists, a truly democratic Germany could only emerge in the framework of political, economic, and cultural integration with the USSR.

Unlike the Americans, the Soviets considered the cultural renewal of Germany essential to the process of political transformation. The Soviets saw culture as a political weapon and considered the work of intellectuals and artists essential to the construction of a new society. In fact, the Soviets had begun planning the organization of the German cultural apparatus even before the fall of Berlin in May 1945. Once SMAD was established, its cultural and political officers—in collaboration with the repatriated German Communists—dominated the German cultural scene. Most of the German Communists exiled in Russia were jailed or murdered by Stalin, accused of being Trotskyists, "conciliators," or Nazi agents. The survivors, hard-core Stalinists and seasoned politicians with extensive contacts in their homeland, returned to Germany with the Red Army. Walter Ulbricht, the former political secretary of the German Communist Party in Berlin and future president of the GDR, began to organize political and cultural networks before the end of the war; his network was later used by SMAD to structure and control German life in the Soviet zone.[28] The Soviets had extensive experience in cultural warfare, which SMAD exploited, launching a massive cultural offensive aimed at attracting German artists and intellectuals. Just two weeks after Soviet troops occupied Berlin, the Soviet military invited a group of Berlin artists to discuss plans for rebuilding cultural life in the city.[29]

Unlike their OMGUS counterparts, the top figures in SMAD were directly involved in cultural policy in occupied Germany. General Sergei Tiul'panov, General Clay's counterpart in the Soviet zone and sector, and Lieutenant Colonel Alexander L. Dymschitz, the head of the cultural division of SMAD's Office of Information, were political cadres in military intelligence who had been selected for their in-depth knowledge of German history, politics, and culture. General Tiul'panov was an economist who spoke perfect German. During World War II he joined the Red Army military intelligence and specialized in the reindoctrination (*retournement*) of high officers of the Wehrmacht and the Luftwaffe who had been taken prisoners in the Eastern front. In 1945 Tiul'panov was sent to Germany, and he became personally involved in day-to-day German politics. He was the creator of the Sozialistische Einheitspartei Deutschlands (Socialist Unity Party of Germany, SED) and supervised propaganda, censorship, and "cultural enlightenment" in the Soviet zone. Dymschitz, a Stalingrad veteran, was Tiul'panov's main operator in the cultural field. Born in St. Petersburg into a highly educated Jewish bourgeois family well connected to literary circles, Dymschitz was an expert on German literature trained at Leningrad State University and the Russian Institute of Literature. During World War II he joined the Red Army, and immediately after the war was sent to Berlin to direct the cultural section of SMAD's newspaper, *Tägliche Rundschau*. In September 1945 he was appointed head of SMAD's cultural section and personally shaped Soviet cultural policy in occupied Germany.

Tiul'panov and Dymschitz developed overt and covert agendas to attract intellectuals and artists. Following the classic Communist strategy, SMAD began organizing "mass organizations" and bureaucratic institutions, supposedly ecumenical but in fact directly or indirectly controlled by Soviet intelligence.[30] In June 1945 SMAD established the Kammer der Kunstschaffenden (Chamber of Art Workers) and a month later created two new organizations to supervise cultural and artistic life in the Soviet zone and sector: the Deutsche Verwaltung für Volksbildung (German Agency for Popular Enlightenment) and the Kulturbund zur demokratischen Erneuerung Deutschlands (Cultural League for the Democratic Revival of Germany). The Deutsche Verwaltung was overtly controlled by SMAD, while the Kulturbund, allegedly "apolitical," was secretly supervised by SMAD's Abteilung für Volksbildung (Section of Popular Enlightenment) and Office of Information.[31]

Soviet political officers viewed all aspects of culture—the press, the fine arts, theater, music, and literature—as resources for strategic propaganda, and did not see a sharp separation between "high culture" and "low culture." From the very beginning of the occupation, the Soviets carried out a well-publicized discussion on the political role of literature, theater, and the fine arts in the reconstruction of Germany. They were also more active than the other three Allies in designing and implementing cultural policy initiatives. SMAD organized large exhibits of German and Soviet art. They were the first of the Allies to reopen theaters and movie theaters in July 1945, showing Soviet films and a selection of German feature films. And they were the first to organize the postwar film industry in Germany, giving German directors and actors the opportunity to produce their own feature films.

At first the content of Soviet cultural policy in occupied Germany was ambivalent, reflecting the conflict between two political undercurrents whose advocates vied for the control of SMAD. There was a public façade of moderation, an insistence that the USSR had no desire to force its political system on Germany, and that the development of German postwar culture was to be left up to the German intelligentsia. At the same time, Tiul'panov carried out a deliberately aggressive agenda that reflected Stalin's plans for a new German culture dominated by the Soviet Union. By the end of 1945 SMAD and the German Communists were calling for the development of a "progressive" culture, under the leadership of the working class, and of a truly "democratic" and realist German culture in the service of social justice.

In August 1946 events in the USSR changed Soviet cultural policy in Germany. The Central Committee of the Soviet Communist Party issued the first of four decrees condemning Western and "cosmopolitan" (that is, Jewish) influences on Soviet literature, cinema, music, theater, and fine arts. This Stalinist cultural offensive, organized and led by Andrei Zhdanov, the third secretary of the Soviet Communist Party, defined an antimodernist line that would be followed by communist parties around the world. Although the Zhdanov decrees were not published by the SMAD-controlled media in the Soviet zone, Dymschitz was instructed to publish a series of articles in the *Tägliche Rundschau* explaining the new Soviet policy and justifying the suppression of modern art in the Soviet zone. By the end of 1946 SMAD had tightened its control over German culture and elaborated an anti-American propaganda agenda against "Western decadence" and capitalism.

American control of information, 1947–1949

In 1947, the Soviet Union consolidated its hold over Eastern Europe and initiated a political, cultural, and ideological offensive against the West. Stalin moved to establish absolute control over communist parties all over the world, subordinating their interests and their policies—including their cultural agendas—to the Soviet agenda. President Truman responded with Executive Order 9835, signed on March 25, 1947, which put into effect the Federal Employee Loyalty Program and authorized intelligence agencies to investigate the beliefs, political allegiances, and personal lives of federal employees in the United States and abroad. Then, on May 12, Truman asked the U.S. Congress for four hundred million dollars in economic and military assistance for Greece and Turkey, and declared the beginning of a global ideological war against communism. In July, secretary of state George C. Marshall announced the American plan to promote the economic recovery of Europe (including Germany) and the Soviet Union. The USSR and its first European satellites, Poland and Czechoslovakia, rejected the Marshall Plan, denouncing it as an aggressive ploy to further the aims of American imperialism.

In part as a reaction to global politics and in part as a response to local events, American propaganda in Germany changed. In Germany, Joint Chiefs of Staff Directive 1779 replaced JCS 1067 on July 15, 1947. The new directive gave OMGUS the task of "encouraging *bona fide* democratic efforts, and of prohibiting those activities which would jeopardize genuinely democratic developments."[32] This meant that the American military government was to challenge not only Nazi ideology but also the supposedly democratic model proposed by the USSR. JCS 1779 further instructed OMGUS to participate more actively in German cultural development. American strategic propaganda, previously directed against Nazism, became refocused on anticommunism. By late 1947, with the emergence of the Cold War, OMGUS expanded its censorship to include communism and any other political visions that contradicted the American agenda.

On the cultural front, the tension between OMGUS and SMAD exploded in October 1947. The turning point was a speech by Melvin J. Lasky at the First German Writers' Congress, held in the Kammespiel Theater in east Berlin. Lasky was a twenty-seven-year-old American journalist living in Berlin, who had arrived in Europe as a combat historian for the U.S. Army. He had been a vociferous critic of Soviet policy in Germany since the beginning of the occupation, and in his speech he denounced the Soviet Union as a police state that conspired against freedom and democracy. General Clay, who had previously been critical of Lasky's denunciations of SMAD, now began to see Lasky as a consultant on anti-Soviet propaganda.[33] That same month, Clay launched Operation Talk Back, a mass-media counteroffensive against Soviet anti-American propaganda. From then on, the media in the American zone and sector became an explicit propaganda tool targeting SMAD and the USSR.

The new OMGUS agenda was accompanied by a change in ICD personnel and by the development of an anticommunist propaganda offensive using all media, including the written press, film, and the fine arts. Most of the New Dealers, liberals, and leftists in ICD were replaced by militantly anticommunist Cold War warriors. Simultaneously, OMGUS removed many left-wing Germans from U.S.-licensed newspapers and journals, replacing them with a new breed of anticommunist American and German journalists. The new editors were instructed by OMGUS to denounce communism, SMAD, and the USSR in their publications.[34] Already in August 1947 OMGUS had created the Documentary Film Unit to produce political films offering the American perspective on the Cold War. That same year, OMGUS cultural officers became involved in covert operations to counteract Soviet antimodernist propaganda and attract German modern artists.

By 1948 the leitmotif of the American political message in Germany was the new friendship between a democratic and prosperous United States and a freedom-loving western Germany. References to Nazi atrocities and German complicity were, if not totally forgotten, relegated to the distant background. The new American propaganda spread a coherent message centered on anticommunism: the United States and western Germany, working as allies, it asserted, would be able to reconstruct Germany and block the expansion of Soviet imperialism. This shift in emphasis allowed the massive reinsertion of former Nazis into German political and economic life, albeit within the rigid ideological framework imposed first by OMGUS and later by the High Commissioner for Germany (HICOG).

The Cold War gave rise to dueling anti-American and anticommunist policies and rhetoric that increasingly precluded any independent political critique. The USSR, SMAD, and the German Communists intensified their anticapitalist propaganda and stressed their interpretation of fascism as the necessary consequence of capitalism. The United States and its allies, they insisted, were rearming Germany and endangering world peace. The Western democracies in turn denounced Soviet authoritarianism and censored anti-American and anticapitalist rhetoric. Censorship, previously directed against expressions of Nazi ideology, expanded to encompass broader targets—capitalism, communism, liberalism, and freedom of artistic expression. As the confrontation between OMGUS and SMAD became more explicit and choleric, ideological boundaries hardened. Both the Americans and the Soviets tried to limit each other's propaganda in their respective zones and sectors, and their tolerance for direct opposition vanished.

Cultural initiatives in the British and the French zones

While OMGUS and SMAD sparred over control of the reconstruction of German culture, the British and the French military governments carried out cultural programs with more limited aims. From the beginning of the occupation, the British Foreign Office instructed the British military government to focus on gaining the confidence of the Germans. The Foreign

Office feared that the American model of denazification would embitter the Germans and could lead to political gains by the Soviets and the German communists. The British, therefore, ended their confrontation policy quickly. The result was a lenient denazification effort, and the British did not radically change the bureaucracy, the universities, and the civil service in their zone.

British cultural policy in Germany often relied on informal contacts at the personal level. Anglo-German discussion groups, clubs, and cultural societies allowed British cultural officers to interact with German journalists, artists, and intellectuals. The paradigmatic example of this approach is the creation of *Der Spiegel.* In 1946 Major John Chaloner, a twenty-two-year-old British information officer who headed the press section in Hannover, and Sergeant Harry Bohrer (originally from Czechoslovakia) had the idea of launching a weekly journal along the lines of *Time* magazine. They named the magazine *Diese Woche,* and selected an editorial committee that included four Germans—among them Rudolf Augstein, a former artillery lieutenant in the Wehrmacht. *Diese Woche* was often critical of the British military government because Chaloner insisted on the importance of giving the German journalists freedom of expression. On January 4, 1947, *Diese Woche* became *Der Spiegel,* and Rudolf Augstein became its co-owner.

On a more formal level, the British military government showed films selected by the Political Intelligence Department of the Foreign Office. These British films projected the "gentleman-ideals" of common sense, fair play, tolerance, and stability. In 1946 the British started Die Brücke, the British information centers that were the forerunners of the British Council Branches. The information centers were primarily libraries meant to reintroduce English literature in Germany. They provided free access to British and international media and included a selection of German books from the Weimar period. The books were also selected by the Political Intelligence Division and, like the films, reflected the middle- and upper-class values of the Edwardian period. At the same time, however, the British Political Intelligence Division was acutely aware of the emergence of the Cold War. While it made available books by authors such as George Orwell and Arthur Koestler, it did not send works by contemporary British leftists to Die Brücke.[35]

The French, the weakest of the Allies, both in military and economic terms, approached the occupation of Germany with a double purpose— to show that France was the cultural hegemon of Europe, and to attract Germans to the culture of Western Europe. French cultural policy was masterminded by the French cultural attaché in Berlin from 1946 to 1948, Félix Lusset. Lusset was a former officer of the French resistance who belonged to the noncommunist left. A fervent antimilitarist, he insisted on the civilian character of France's cultural mission in Germany. This led to confrontations with the French military occupation authorities, and eventually to his dismissal.

Yet Lusset was able to shape an active, varied, and successful cultural program. He wanted to project French culture (*rayonemment culturel*) and revive cultural ties between France and Germany. During his two-year tenure, Lusset organized three memorable cultural events: the visits to Germany of Albert Béguin (November 1946), Jean-Paul Sartre (January–February 1948), and Jean-Marcel Bruller, who published under the pseudonym Vercors (June 1948).[36]

The French military government was also active in the fine arts. As early as 1946, it organized art exhibits, using the French artistic tradition as a symbol of the country's cultural might. Among the exhibits were surveys of recent French painting, printmaking, and graphic art, all in 1946, and ceramics and sculpture ("From Rodin to the Present") in 1947. These exhibits featured the work of modern artists like Modigliani, Picasso, and Chagall. The French also showed classic German art in order to emphasize the essential cultural role played by Germany in Western civilization.

The four occupying powers did more than impose new political, economic, and legal paradigms in Germany. Each military government attempted to project the cultural paradigms of its nation, and to offer credible and positive programs to guide German cultural rebirth. Their psychological warfare specialists sought to replace Nazi imagery with a new set of images, and new conceptions of culture and aesthetics, that were intrinsically tied to their respective ideologies and agendas.

One

American Atrocity Propaganda

In April 1945 the Western Allies liberated Nazi concentration camps in western and southern Germany. The highest echelons of SHAEF—Generals David D. Eisenhower, George Patton, and Omar Bradley among them—went to see the camps soon after liberation. After his first visit to Ohrdruf, the first camp liberated by the Americans, General Eisenhower ordered the military units in its vicinity to tour the camp. Eisenhower also invited members of the U.S. Congress, selected journalists, and a group of Hollywood personalities to see the Nazi atrocities for themselves. At the same time, the field commanders of the U.S. Army forced German civilians living near the camps to visit the sights of genocide. These forced visits inaugurated the confrontation policy, the first American psychological warfare operation in occupied Germany to rely primarily on visual propaganda.[1]

This propaganda policy fit comfortably within the aims and the spirit of American pre-occupation strategic thinking. In December 1944 the Office of War Information insisted that "our primary task is to make *them* [the German population] realize that they are guilty."[2] The Americans used the concentration camps as evidence of Nazi criminality, to justify the war retroactively, and to stress the doctrine of German collective guilt.

The confrontation policy used all available media, but it had a strong visual component—guided forced visits to the liberated concentration camps, booklets and billboards with atrocity photographs, and atrocity films. American field commanders forced Germans living in the vicinity of concentration camps to tour the sites. There were at least twenty-four such "confrontation visits" in the first few days after liberation, although there may have been more.[3] The occupying forces placed improvised poster displays of atrocity photographs in towns and cities. Some of these posters were titled aggressively—"German Culture" or "These Atrocities: Your Guilt."[4] The confrontation policy also used film. *Welt im Film,* the joint Anglo-American newsreel produced in London for occupied Germany, dedicated its fifth issue (released on June 15, 1945) to the camps.[5] And in 1946 the twenty-two-minute documentary *Todesmühlen* ("Death Mills") was shown—on a voluntary basis—for a few weeks. In this chapter, I analyze American atrocity propaganda as a psychological warfare campaign based predominantly on visual methods.

The Nazi camps and American propaganda for the home front

As soon as the SHAEF troops entered Germany in September 1944, the Allied Forces Network and the U.S. Army newspaper *Stars and Stripes* warned the occupying troops against a false sense of familiarity. Allied soldiers were instructed to avoid fraternization with Germans. A typical slogan was "If, in a German town, you bow to a pretty girl, or pat a blond child . . . you bow to Hitler and his reign of blood . . . you caress the ideology that means death and persecution. Don't fraternize."[6] In early 1945, JCS 1067, deeply influenced by Morgenthau's thesis of German collective guilt, stated that Germany "will not be occupied for the purpose of liberation but as a defeated enemy nation" and explicitly banned fraternization with German civilians.[7] American propaganda insisted that the similarities between Americans and Germans were only skin deep, and that they hid irreconcilable moral differences.

Even so, the American troops who invaded Germany were not prepared for the horrors they discovered in the Nazi concentration camps. Ohrdruf was found by accident as units of the Fourth Armored Division of General Patton's Third Army searched for a Nazi communications center. The small labor subcamp of Buchenwald had only a few surviving inmates (from an original population of ten thousand) and mounds of corpses. Soon after, the U.S. Army found even more gruesome camps to the north, near the town of Nordhausen. In the days that followed, a string of concentration camps and labor camps were discovered throughout Germany.[8]

The smell and the sights in the concentration camps were, according to liberator testimonials, unforgettable. Leonard Linton, a paratrooper from the 82nd Division with extensive combat experience, recalls that nothing he had seen either in training or in battle had prepared him for his entry into the Wöbellin concentration camp, a subcamp of Neungamme. Years later, he recalled that "seeing and smelling the cadavers left a much more powerful mark on us than just reading about other camps that were liberated shortly before."[9]

On April 12, a week after the liberation of Ohrdruf, Generals Eisenhower, Patton, and Bradley toured the camp. They saw the piles of corpses and examined the torture devices used by the SS. The sights and smells of the camp, and the testimonies of inmates, shocked the American generals.[10] General Eisenhower ordered every U.S. military unit in the vicinity to tour the camp in order to reinforce their sense of mission. After this first use of Ohrdruf as explicit visual propaganda, the concentration camp became a central trope in American media targeted at GIs. American propaganda exploited the atrocities to emphasize the alleged difference between Americans and Germans and to reinforce the nonfraternization order. In October 1945 the U.S. Army weekly *Yank* published an editorial explaining that "a concentration camp cancels a clean bathroom and attempted mass extermination of a race overbalances a sunny disposition."[11] "This Is Why We Fight," a "newsmap" prepared and distributed by the U.S. Army Signal Corps, insisted that the visual evidence of the Nazi camps was proof of German evil. The pamphlet included photographs of corpses, the image of an emaciated inmate, and a caption that made the point explicitly:

Now we've SEEN it. This is what fascism's "New Order" brought to millions in Europe—death by torture, starvation, flogging, and every fiendish method the twisted German mind could devise. Not so long ago some of us were saying "the Germans are really nice people, pretty much like us—decent, clean, and kindly at heart." These are the sights—photographed by Signal Corps cameramen—which greeted our soldiers who over-ran German concentration camps.[12]

The concentration camp thus became the symbol of German depravity.

This was a dramatic shift in American anti-Nazi propaganda. During the war, reports and rumors about the existence of Nazi camps had been met with disbelief in both the United States and Europe.[13] In World War I, American propaganda had accused the Germans of horrific crimes that never actually took place, and the Roosevelt administration was careful not to repeat that blunder. Although the administration knew of the existence of German concentration and extermination camps as early as 1943, genocidal anti-Semitism was not part of the American narrative of Nazism. Rather, American wartime propaganda portrayed Germany as a ruthless and expansionist aggressor, and the *Führer* as a charismatic madman commanding a delinquent gang, who had suppressed democratic freedoms and imposed a regime of terror in his own country and in occupied Europe. The characterization of the German population was more ambiguous. Sometimes Germans were described as automatons who followed Hitler blindly. Other times, distinctions were made between the Nazis and the "good" Germans.

Even after the Red Army liberated Majdanek, on July 23, 1944, and Auschwitz-Birkenau, on January 27, 1945, few American or British journals showed the images recorded by the Soviet photographers and camera teams. *Life* and *Illustrated London News* did publish some photographs of Majdanek—building exteriors, Zyklon B containers, piles of passports, luggage locks, and shoes—but no images of victims or crematoria were released. In April 1945 the American public received early reports about Nazi atrocities with skepticism. On May 5, as the liberation of the camps continued, a Gallup poll asked Americans if they believed "reports that the Germans have killed many people in concentration camps or let them starve to death." Only 40 percent said they did; 52 percent did not answer or said that they did not know.[14] Even American soldiers tended to be skeptical. Private John McKisin of the Seventh Army, one of the liberators of Dachau, told a reporter that until he arrived at the camp, he had dismissed reports of German atrocities as propaganda.[15]

After the liberation of the concentration camps in Germany, General Eisenhower decided that the home front should also be exposed to the visual evidence of Nazi atrocities. The written accounts appearing in the American media were inadequate, he thought; to imagine the reality of the camps it was necessary to *see* them. He asked that prominent American politicians and reporters travel to Germany to look at the camps themselves. He hoped that these eyewitnesses would be persuaded that the Germans had committed unimaginable crimes and would then transmit their impressions to the home front. On April 19, Eisenhower sent a cable to General George C. Marshall:

> We are constantly finding German camps in which they have placed political prisoners where unspeakable conditions exist. From my own personal observation, I can state unequivocally that all written statements up to now do not paint the full horrors. In view of these facts, you may think it advisable to invite about 12 congressional leaders and 12 leading editors to see the camps. If so, I shall be glad to take these groups to one of these camps. Such a visit will show them without any trace of doubt the full evidence of the cruelty practiced by the Nazis in such places as normal procedure.[16]

President Harry S. Truman, who had succeeded President Roosevelt on April 12, was shown film footage of the liberation of the camps and authorized the visits.

Twelve members of the U.S. Congress, six Republicans and six Democrats, flew to Germany with the specific purpose of touring the concentration camps to view the evidence.[17] The bipartisan committee visited Buchenwald in Saxony, Nordhausen in Thüringen, and Dachau in Bavaria. On April 24, 1945, the committee arrived in Buchenwald accompanied by Brigadier General John M. Weir, Colonels Robert H. Thompson and John A. Hall, and a group of photographers. They toured Dora at Nordhausen on May 1 and Dachau on May 2—a scant forty-eight hours after the first American troops entered the camp.

Concurrently but separately, a group of eighteen American newspaper and magazine editors toured the camps.[18] The group visited Buchenwald on April 25, only twelve days after the camp had been liberated by the Third Army, and later went to Dachau. Joseph Pulitzer, the publisher and editor of the *St. Louis Post-Dispatch,* who headed the delegation, later wrote that seeing the concentration camps radically transformed his understanding of Nazism: "I came here in a suspicious frame of mind, feeling that I would find that many of the terrible reports that have been printed in the United States before I left were exaggerations, and largely propaganda." Seeing the camps proved that he had been mistaken; the previous reports had been "understatements."[19] Pulitzer had not imagined the degree and magnitude of the Nazi crimes. Only after seeing the camps himself was he able to grasp their immensity.

After the delegations' visits, Nazi concentration camps became the most talked-about news in the United States. German war crimes were discussed on the radio, and American newspapers and journals, both military and civilian, published accounts of Nazi atrocities. Photographs most often took the place of lengthy narrative descriptions. In April 1945 an article in *Stars and Stripes* claimed that no medium except photography could bring "home to a civilized world . . . the cold truth of German cruelty and sadism." *Life, Newsweek, Illustrated,* and *Time* all published photographic spreads showing the camps during the first week of May. *Vogue* published Lee Miller's photos of Buchenwald in June 1945, under the title "Believe It!" "There are no words in English which can adequately describe the Konszentrations-Lager at Dachau," read the caption of a photograph showing a pile of corpses covering the entire page. Photographs in other publications had captions such as "Seeing Is Believing," "The Pictures Don't Lie," "Indisputable Proof," and "This Is the Evidence." Images made the stories of Nazi atrocities believable, and became the main elements used to imagine the Holocaust.[20]

In June 1945 Pulitzer, in collaboration with the U.S. government, organized *Lest We Forget,* an exhibit of atrocity photographs. Pulitzer was convinced that photography was the ideal way to provide the American public "true evidence of what went on." The show opened in St. Louis with twenty-five twelve-foot-high photo murals, enlargements of photographs taken at the camps by the Army Signal Corps, the Associated Press, and the British Army. No admission was charged. The exhibit also included two films. One comprised an hour of uncut and unedited footage shot by the Signal Corps documentary unit. The other was Frank Capra's *Know Your Job in Germany,* a piece produced by the Signal Corps to persuade American soldiers of the evil of the German enemy. A total of 81,500 people saw the free showings in the Kiel Auditorium Opera House in St. Louis. The popularity of the films proved so great that there were forty-four screenings, rather than the twelve initially programmed. The St. Louis show was seen by 4,919 people on the opening day, and a total of 80,413 people visited the exhibit in twenty-five days.[21]

The exhibit then traveled to Washington, D.C., where it opened at the Library of Congress on June 30 under the auspices of the *Washington Evening Star* and the *St. Louis Post-Dispatch*. Senators Alben Barkley, Walter George, and Leverett Saltonstall, Representative Dewey Short, and Benjamin M. McKelway, associate editor of the *Washington Star,* all of whom had toured the concentration camps, spoke at the opening. Close to ninety thousand people attended the exhibit in Washington during its four-week duration, more than had visited any previous Library of Congress exhibit. Chief librarian Luther Evans commented that the photomurals and films "represented a more unanswerable indictment of the enemy's sadistic brutality than could any aggregate of written documents and personal reminiscence." The atrocity photographs subsequently toured Boston, Cleveland, and New York. Demand was so great that the *Post-Dispatch* made a duplicate set of the photographs to tour the Midwest, and eventually a third set was created and circulated.[22]

The exhibit highlights the effectiveness of the photographic image in mass education and political propaganda. Like words and sounds, images are vehicles of meaning and instruments of memory. They move and excite, simplify and exemplify, instruct and persuade. And photographs—although they can be staged, retouched, reiterated, or presented subliminally to maximize their effect—are seen as truthful and authentic. Susan Sontag writes that photographs have the "credentials of objectivity [because they are seen as lacking] the taint of artistry, which is equated with insincerity or merely contrivance." If photographs are sponsored by a reliable authority, they tend to be accepted as incontrovertible evidence—if an event has been photographed, it must have happened. The viewers of photographs become what Jennifer Mnookin terms *"virtual witnesses."*[23]

Lest We Forget can be inscribed in a long tradition of political photography that dates back to the 1850s. In the mid-nineteenth century, European nation-states scrambled to exploit African natural resources, and this led to the imposition of increasingly brutal forced labor systems. The British Evangelical movement, in denouncing forced labor and atrocities in Africa, exploited the propaganda opportunities afforded by the new photographic technology. The missionaries used atrocity photographs to make visible the suffering of Africa to British audiences. The development of the Kodak camera in 1888 and advances in printing technology allowed the mass reproduction of atrocity photographs in the popular press, and the innovative lantern-slide lectures. From then on, photographs became fundamental documentary evidence to prove atrocities and war crimes.[24]

The British example was soon replicated in the United States. In the 1860s photographs were used in antislavery propaganda, in fund-raising for abolitionist causes, and in Civil War propaganda. In the early 1900s Lewis Hine's photographs of child workers contributed to the passage of new labor laws. Photographs were also used to push for environmental legislation. And in the 1930s the photographers of the New Deal recorded the social ills plaguing the country and the consequences of bad land management and abusive agricultural practices.[25]

Moreover, photography revolutionized the depiction and perception of war. During the Crimean War (1854–1856), the British government commissioned Roger Fenton to record the conflict. Fenton was instructed to shoot his photographs so as to create a positive impression of a war that was unpopular with the British public. The War Office ordered him not to photograph the horrors and chaos of war but to make warfare appear clean, stately, ordered, and devoid of casualties. During World War I, by contrast, the camera captured the extent of the mass carnage better than words could. Even as the governments involved in the conflict censored the media to avoid publication of gruesome pictures, pacifist activists used photographs and film to illustrate the monstrosity of warfare. Photojournalism came into its own during the Spanish Civil War (1936–1939), and continued to develop throughout World War II. Censorship continued, but war photographs published in magazines like *Life* and *Look* in the United States, *Picture Post* in the UK, and *Signal* in Germany brought the battles to the home front, allowing civilians to join the fight in their imaginations.[26]

The photographs of the Nazi camps published in 1945 differed drastically from earlier images of war provided to the American public. Americans had experienced World War II through censored images. After Pearl Harbor, the Office of Censorship was set up to control the information and images that reached the American public. The War Department's Bureau of Public Relations also screened photographic material and determined what was appropriate for publication. The aim was to maintain public support for the war effort, and the fear was that images of defeat or suffering would demoralize the public. By 1943, however, the U.S. government worried that the censorship of images was causing skepticism and disbelief. From then on, carefully selected photos of dead soldiers were released, but censors still blocked the publication of photographs showing decapitated or dismembered American soldiers. Images that were deemed potentially upsetting to the viewer—such as photographs of civilian victims—were not shown at all. The combination of government censorship and the self-censorship of the press resulted in a cleansed and orderly image of war and death.

Even images of death in combat, however, contrasted sharply with the atrocity photographs shown in "Least We Forget" and in the American press. The American public had never before been exposed to such grisly subjects: masses of unburied, emaciated corpses, often naked, reduced to living skeletons. The photographs were radically new, yet they built on a preexisting propaganda campaign. In testifying to German moral depravity, they advanced a rhetoric of American moral superiority and helped consolidate the idea of World War II as the "good war."

The Nazi camps and American propaganda in Germany

American psychological warfare in Germany exploited the revelations of the Nazi camps to expose the evilness of Nazism, to prove the collective guilt of the German people, and to establish the moral superiority of the Allies. Immediately after the discovery and the liberation of the concentration camps in central and south Germany, the U.S. Army launched its confrontation policy. The policy seems to have had five aims. First, it forced the German public to witness the evidence of Nazi criminality. Second, it challenged the Nazi narrative of the war. Third, it attempted to make Germans feel responsible for abetting the Nazi regime. Fourth, it tried to replace feelings of victimhood with a sense of culpability. And fifth, it made Germans recognize the new structure of power. Germans were defeated, occupied, and subject to the whims of the occupying powers.

American soldiers were baffled when civilians living nearby claimed not to have known of the existence of the concentration camps. J. Paul Heineman, an American GI involved in the liberation of Landsberg (a Dachau subcamp), recalls that German civilians "when asked why they didn't do something to stop the killings," replied, "Oh, we didn't know what was going on out here!" Ed Granland, another liberator of Landsberg, remembers his shock at the denials:

> I have never in my life seen such skinny people as those inmates were. At the same time we talked to some civilians. They lived within a mile of the camp. They denied knowing anything that was going on, even with the strong smell of human flesh. This was the worst day of the war for me.[27]

These were not isolated cases of denial. Photographer Margaret Bourke-White described the incessant denials of knowledge:

> "We didn't know! We didn't know!" I first heard these words on a sunny afternoon in mid-April, 1945. They were repeated so often during the weeks to come, and all of us heard them with such monotonous frequency, that we came to regard them as a kind of national chant for Germany.[28]

An April 1945 U.S. Army report on the conditions of Buchenwald at the time of liberation, summarized the official American interpretation of the German claims of ignorance.

> Most Germans claim, of course, that they knew little or nothing about what went on in such camps as Buchenwald or Ohrdruf or Auschwitz.... A large number of the Germans now being interrogated are indeed lying.... No normal German living in the area of Weimar or Gotha could fail to have a fairly clear picture of the general proceedings at Buchenwald and Ohrdruf. The stories of these camps were told over and over among Germans by the guards or by the German civilians who worked in the camp and returned home each night.... The secret of blissful ignorance lies in the mental block which most Germans seem to have set up, to exclude from their daily consciousness knowledge of this sort. But however deeply buried in the minds of individuals, a certain essential knowledge of what has happened remains. Probing by Allied interrogators almost invariably reveals this to be the case.[29]

The confrontation policy began with forced tours of the camps. American atrocity propaganda made the concentration camps visible and undeniable. The Americans wanted Germans, irrespective of gender, age, and social status, to see the camps. They regarded all Germans as accomplices of the Nazi criminal project—a hypothesis of collective guilt with a corollary concept of collective punishment. The camp visits were regimented, supervised, and mandatory. German civilians were brought to the camps in groups, under military escort. Once inside, they were forced to see, touch, and smell. Some German civilians were required to dig mass graves, transport corpses out of the camp, and bury bodies. On April 15, 1945, General Patton ordered the mayor of Weimar to gather a thousand inhabitants of the city, half of them women, to tour Buchenwald, the local concentration camp. As he later recalled, fifteen hundred Germans marched through the camp and were "made to look at the horrid spectacle."[30] It was a six-hour excursion that started with a march to the camp and continued with a tour under the supervision of American soldiers. The civilians had to view the piles of corpses and to look into the open graves. They were taken through the barracks to witness the conditions in which human beings had been forced to live and made to examine the torture devices used by the SS. Joseph Pulitzer, in his booklet *A Report to the American People,* described the scene as it was portrayed in a documentary film:

At Buchenwald, site of one of the most horrifying of Nazi prisons and torture camps . . . residents of the community are shown trudging along the highway toward the camp they have been ordered to inspect. Most of them are marching gaily, many smiling and waving as though they were going to a picnic. Only a few had the decency to cover faces as they approached the Signal Corps cameraman. They are shown later amidst the gruesome remnants of the camp. Then the camera is turned to register expressions. Their faces are twisted with revulsion at what they have seen.[31]

Similar forced viewings occurred in other camps. Wöbellin, in the outskirts of Ludwigslust, was liberated by units of the 82nd Airborne Division. General James Gavin, the division commander, ordered all residents of the town older than ten to visit the camp, where they were shown around by armed GIs. Lieutenant Colonel Edward F. Seiller, one of the first officers to enter Kaufering Lager No. 4 near Landsberg, ordered that a group of 250 civilians, including ministers, priests, farmers, businessmen, and common laborers from the surrounding farms, be brought to see the bodies of the dead inmates killed by torture. He too "wanted them to see for themselves."[32]

American field commanders thus transformed the sites of slave labor and murder into didactic museums. Curators of museum exhibits select and organize objects, complemented with relevant historical, socioeconomic, or aesthetic data, in concentrated displays that assert a certain narrative of past events. Once the geography of an installation is established, it may be very difficult to see its elements in any other way. Similarly, the camps, as seen in these forced visits, were made to serve an educational agenda—the exhibition of the German criminal project. The camps and the corpses, the places of brutality and the victims of brutality, were placed on display.

The concentration camp was not, of course, a typical museum. Attendance at museums is, in general, voluntary. Visitors are free to choose their paths through an exhibit, to skip over pieces according to their tastes and interests, to accept or ignore the thesis suggested by the curator. This element of choice was absent in a confrontation tour. German civilians living near the camps were forced to participate and, guided by GIs in combat gear, to follow a fixed path through barracks and open mass graves. An aspect of the museum as educational institution was nonetheless evident in the arrangement of the relics of Nazi atrocities according to a plan. At Buchenwald, for instance, SS mementos of obscene gruesomeness—pickled human organs, remnants of human skin, shrunken human heads—were placed together on a table and displayed to viewers—be they American delegations or German civilians. A mannequin was hung in a noose to demonstrate how one common torture device functioned.[33]

Corpses were central elements in the visual displays. The piles of bodies on carts were not always those left by the SS. One of the most famous photographs of Buchenwald—a wagon overflowing with emaciated corpses—belonged to an exhibit in preparation. Since the bodies withered and disintegrated, every few days the GIs replaced the older corpses with newer bodies, reconstructing the pile for the new confrontation tours. The evidence was restaged to create a visual narrative of Nazi atrocities.[34]

The American military also organized collective burials for the dead inmates and forced German civilians to participate. In Ludwigslust, corpses were loaded in open carts and carried to the town center, where a permanent cemetery was created in the central square, in front of the town hall. Public burials were also held in Hagenow and Schwerin, with German civilians ordered to visit the burial site before the graves were closed. Civilians in Nordhausen were forced to bury the bodies of the victims. Brigadier General Paul Adams had a cemetery and memorial constructed in

Dachau.³⁵ These highly visible burials had symbolic meaning and political significance. While the Nazis had used ghettos and concentration camps to hide their victims, making the horror invisible and therefore unknowable, the Americans made the corpses and the atrocities the symbols of the Third Reich, immediately accessible to public consciousness. In effect, the American soldiers transformed the camps into patrimonial sites, the definitive *lieus de memorie* of the Third Reich.

The confrontation visits and the burials had a characteristic rhetoric. Often speaking through interpreters, the American commanding officers addressed the German civilians and held them responsible for the Nazi atrocities. At the Wöbellin burials, Colonel Harry Cain insisted on German collective guilt:

> Untold numbers of other Allies soldiers and German citizens shudder before similar burial services as you shudder now. The Allies shudder because they never dreamed or visualized that human leadership, supported by the masses, could debase itself as to be responsible for results like those who lie in these open graves. You Germans shudder for reasons of your own. Some of you having been a party to this degradation of mankind, shudder for fear that your guilt will be determined, as in fact it will. Others among you shudder because you let depravity of this character develop while you stood still and did nothing about it. The civilized world shudders on finding that a small part of it's [sic] society has fallen so low. . . . That world must, as it does, hold the German people responsible for what has taken place within the confines of this nation.³⁶

Cain's speech focused on the alleged difference between Americans and Germans, and between the "civilized world" and Germany. In Kaufering, Lieutenant Colonel Seiller told German civilians that they shared the blame for Nazi atrocities: "You may say that you weren't personally responsible for all this, but remember you stood for the government which perpetrated atrocities like these."³⁷

The initial impact of atrocity propaganda in Germany

The confrontation policy was based on the assumption that punishment and recrimination would induce a sense of guilt. The German population, it was thought, would react to the visual evidence of Nazi atrocities with shame. But less than three months after the liberation of Ohrdruf,

SHAEF's Psychological Warfare Division (PWD/SHAEF) issued a confidential report that challenged these assumptions. *Atrocities: A Study of German Reactions,* presented an analysis of Germans' attitudes toward "the problem of atrocities" based on PWD field research and the study of responses to a booklet of atrocity photographs. The booklet, *KZ: A Pictorial Report from Five Concentration Camps,* was prepared by the OWI using photographs from the liberation of Buchenwald, Bergen-Belsen, Gardelegen, Nordhausen, and Ohrdruf and was "pretested"—shown to one hundred German civilians—in Cologne, Kassel, Erfurt, Koblenz, Kaiserslautern, Marburg, Heidelberg, and a number of small villages. Later, a "test sale" of two thousand copies of *KZ* was conducted in Heidelberg and Kaiserslautern.[38]

PWD/SHAEF determined that atrocity propaganda had failed to create "feelings of guilt." According to *Atrocities,* "almost every German has had some contact with our campaign to expose the facts of German concentration camps." Over half of the interrogated subjects believed in the veracity of "Allied propaganda" on the camps, and only 15 percent doubted the extent of the atrocities. Yet most Germans were still unclear of the "details"—number of camps, number of victims, and living conditions in the camps. When asked to estimate the number of inmates killed in Nazi camps, most of the Germans guessed "tens of thousands." Furthermore, the vast majority of the German civilians interrogated believed that responsibility for the atrocities revealed lay exclusively with the Nazi Party and the SS. They claimed that Nazi terror and propaganda prevented people from knowing the truth and reacting accordingly. The report quotes a typical answer:

You Americans can hardly understand the conditions under which
we were living. It was as if all of Germany were a concentration camp
and we were occupied by a foreign power. We were unable to do anything to oppose them. What could one person do against the powerful
organization.

This argument, the report said, was presented "with automatic regularity" by Germans "of all types and social classes." Many considered atrocities "the inevitable consequences of war."[39]

German response to the atrocity booklet proved to be complex. The test sales of *KZ* in Heidelberg and Kaiserslautern were a success—all copies were bought in less than an hour—but from a propaganda point of view, it was a failure. Germans took *KZ* as evidence of Nazi depravity but not as a more general indictment of German behavior during the Third Reich. The photographs elicited "what appeared to be genuine reactions of deep horror and shame." Yet "no feelings of guilt were produced among any of the readers; when the sense of shame was noted it was a purely personal reaction without any feeling of co-responsibility for allowing atrocities to have taken place."[40]

While the *KZ* photographs had more impact than the text, the images did not surprise the viewers in the way that PWD/SHAEF had expected. In fact, respondents compared American and Nazi propaganda. Many of the Germans interviewed claimed that the photographs in *KZ* made them "recall the Nazi booklets about Katyn," the 1940 massacre of thousands of Polish prisoners of war by the Soviet Red Army. In 1943 Nazi propaganda minister Joseph Goebbels had made Katyn the focus of an anti-Bolshevik and anti-Semitic multimedia campaign. The propaganda booklet *Amtliches Material zum Massenmord von Katyn* was illustrated with photographs taken at the site of the massacre. Germany's leading movie studio, Universum Film AG (Ufa), released an atrocity film, *Im Wald von Katyn,* that was shown in all major German cinemas. In sum, the German public had been exposed to Nazi atrocity propaganda and reacted with skepticism to the American version, which seemed remarkably similar. PDW/SHAEF found that educated readers thought the Western Allies were "overplaying atrocities."[41]

Atrocities shows that the American and the British intelligence communities were already, in June 1945, worried about the impact of Soviet propaganda on the German public. PWD/SHAEF realized that the confrontation policy invited negative comparison with the Soviet propaganda line. In spite of the brutality of the Soviet takeover of Berlin, by June the Red Army was already involved in the process of winning the hearts and minds of German civilians. While the Western Allies operated on the premise of German collective guilt, the Soviets stressed that the German "people" had been innocent victims of a fascist dictatorship. International monopoly capitalism, German militarism, and German big industry were blamed for terrorizing the "good Germans" and forcing them into submission. Concentrating blame on the class enemy while ignoring the behavior of the overall population under Nazism had the great political advantage of not alienating the Germans. In fact, the Soviets were combining harsh punitive measures with a positive message that echoed Stalin's 1942 assertion—"Hitlers come and go, but the German people, the German nation, remains."[42]

In this context, it is not surprising that the Germans interrogated by PWD/SHAEF warned that the Soviets were winning the propaganda war:

> Both anti-Nazis of the non-Communist variety and middle-class conservative or nationalist elements, reacted to the KZ booklet and to our entire atrocity propaganda by citing in contrast the output of the Russian-sponsored Radio Berlin, which, just at the time of these interrogations, was taking a non-hostile attitude toward the German people.... They usually started out by saying that BBC and [Radio] Luxembourg constantly harp on atrocities and hold all Germans responsible, while Radio Berlin was not so "unfriendly" to the German people in that it drew a distinction between the guilty and those who merely stood by. They usually added what they believed to be other marked differences in the propaganda approach of the West and the East, namely that the Russians told of food being brought into Berlin and Dresden, while the Allies were emphasizing that the Germans would have to work hard this fall or starve. The Russian radio spoke of fraternization and good will between the Russian Army of Occupation and the citizens of Berlin, while the Western Allies are conspicuously quiet on the subject.... The propaganda implication in their minds was clear: the Western Allies must stop stressing collective guilt if they hoped to counteract Russian influence.[43]

In a political context charged by the rekindled competition between the United States and the USSR, the confrontation policy was not only ineffectual but risky.

The evolution of atrocity propaganda

Only a few thousand Germans were exposed to Nazi atrocities through the forced tours. The confrontation visits lasted approximately a month after the liberation of the camps. On May 4, 1945, General Eisenhower asked Washington to stop the visits of American VIPs, and on May 9, General Omar Bradley, the commander of the U.S. 12th Army Group, sent a secret message to Eisenhower recommending an end to the confrontation tours. The camps had been cleaned up and the corpses buried. No evidence of atrocities remained, according to Bradley, and therefore no educational value could be derived from the visits. Indeed, in their current state, the camps might reinforce skepticism about the existence of Nazi atrocities.

After May 14 the only visitors allowed were Allied technical and medical teams collecting evidence for the war crime trials.[44]

Yet photographs and objects from the camps were used as legal evidence. On November 29, 1945, during the proceedings of the International Military Tribunal in Nuremberg, the prosecution introduced an hour-long film titled *The Nazi Concentration Camps* as visual evidence of Nazi atrocities. Lawyers complemented oral arguments with a complex system of visual exhibits—photographs, objects, diagrams, models—making the courtroom a mini-museum, where visual evidence was displayed and interpreted. Film and photography were accepted as the ultimate witnesses—impartial, truthful, unaffected by prejudice and errors of memory.

While continuing to use the camps as evidence of Nazi criminality in a judicial context, the Americans changed their propaganda strategy in Germany. On November 9, 1945, just six months after V-E Day, Byron Price, special advisor to Generals Eisenhower and Clay, cautioned that American propaganda had to develop a more forward-looking message:

> Our propaganda needs to be given an increasingly positive character, in contrast to the long-continued attempt to impress the Germans of their collective guilt, which from now on will do more harm than good. A story circulates among the Germans to the effect that one radio listener who followed the Allied broadcasts throughout the war because they gave him hope, has now put away his receiver because he hears only condemnation and abuse.[45]

The campaign that highlighted German collective guilt was not attracting the German population. From a propaganda standpoint, this is not surprising. In order to avoid resistance and rejection, strategic propaganda must not antagonize its target.

Moreover, the collective guilt hypothesis had never been universally popular within the War Department, which oversaw OMGUS, or in OMGUS itself. Many American career officers had strong links to Germany and admired the professionalism of the Wehrmacht, and some even considered themselves Germanophiles. Many also rejected the genetic/psychological hypothesis of German evilness, and considered Nazism as a political phenomenon triggered and sustained by the contradictions of modern capitalism (a conclusion oddly consistent with the Soviet line).[46] They were committed to collaborating with their German peers to build a democratic, pluralist society. On October 1, 1945, the U.S. nonfraternization policy officially ended. GIs were allowed to speak to Germans and, a few months later, were encouraged to associate with Germans as "ambassadors of democracy."[47]

Increased contact between American soldiers and German civilians—especially women—further undermined the American position on the issue of German collective guilt and complicated the official distinction between conqueror and conquered. In October 1946 there were seven million more women than men in occupied Germany, and the demographic imbalance was particularly stark in the western zones. At first liaisons between American army personnel and German women were illegal and punishable by court martial and fines. As the months passed, however, relationships between American soldiers—often new recruits who had not fought against the Nazi army—and German women, became more common.[48]

These interactions changed the dynamics of occupation, chipping away at the image of Germans as alien and as the enemy. A U.S. Army survey taken in September 1945 showed that four out of ten American soldiers in Germany had a "fairly favorable opinion of the German people." Moreover, 57 percent did not blame the Germans as a whole for starting the war, and only 25 percent blamed German civilians for the concentration camp atrocities. Many of the soldiers stated that they had developed a more favorable opinion of the Germans since arriving in Germany. While soldiers disliked the Germans' "air of superiority and arrogance," they liked their "cleanliness and industriousness," and most admitted having had social interactions with German women.[49]

By 1946 American rhetoric concerning German guilt had changed. Nazi atrocities were attributed to individual perpetrators. The German "people" were exonerated from direct responsibility for the criminal deeds of the Third Reich. The system of denazification implied that Germans could be sorted according to their different degrees of complicity with the Nazi regime—a departure from the notion of collective guilt. The concept of individual guilt was reinforced by the 1946 Nuremberg trials. Justice Robert H. Jackson, the chief counsel for the Allied prosecution, was adamant: "We have no purpose to incriminate the whole German people."[50]

American propaganda tailored to American soldiers, however, continued to use the concentration camp as a marker of Germanness. *Occupation*, a 1946 handbook issued by U.S. Forces, European Theater (USFET), warned newly arriving American personnel of stark differences between Americans and Germans:

After 6 years of propaganda, you are going to be surprised when you see your first Germans. We have talked so much about them, hated them so much, and read so much, that we are apt to think of them as different from other people. They are, but not in a way that you can see. Just as German cities are apt to remind you of America, so will the people remind you of Americans. . . . That's why you have to keep in mind what the German people are like. Central heating is typical of Germany, but so was Buchenwald where mass murder was performed with typical German efficiency. German cleanliness is typical—so much so that they tried to make soap out of human bodies. Nazi art gave the world lampshades of decorated human skin. This isn't being like Americans![51]

USFET wanted to make sure soldiers who had not fought in World War II were aware of the recent German past:

Coming to Europe at this time in one way puts you behind the eight-ball. You are seeing half of the story, the other, the worst half, is buried. Lidice is an open, grassy area with a simple cross to mark the place where the Nazis wiped out a whole Czechoslovakian town. The sites of Buchenwald, Dachau, and Belsen are cleaned up, and the starved, scorched bodies have been recently buried.[52]

Until early 1947, the official American literature still proffered the concentration camps as the symbol of Germany and the defining trait of Germanness.

Of course, the emergence of the Cold War further changed American policy toward Germany and the Germans. As early as 1947, the U.S. government considered the American zone and sector as Western Germany, an ally of the anticommunist West. The westernization of the American zone and sector pivoted on the notion that Germans were, like Americans, a democracy-loving people. Western Germans were now seen as citizens willing to create a new independent, liberal, and capitalist country fully integrated—economically, politically, and culturally—into the Western alliance. This radical change in perception necessarily deactivated the confrontation policy and the thesis of German collective guilt.

The Cold War shaped an American foreign policy that increasingly relied on the U.S.–Western German anticommunist alliance. The concentration camp disappeared from American propaganda in Germany, and Nazi atrocities receded to the distant background. The American efforts in Dachau in 1951 are emblematic.[53] In order to "bridge the sea of misunderstanding," the U.S. Army built a community center in Dachau, and American officers invited local notables to a New Year's party at the officers' club.

American officers organized a community-wide Christmas fund, encouraging civic cooperation between Catholics and "Evangelicalists" [sic], and tried to integrate the American military into the community. An ice rink in the U.S. Service Center, previously reserved for American children, was opened to Germans. The center's new kindergarten likewise accepted German children to foster "a spirit of comradeship between children of the two nations." "Prejudice has no place on Dachau's playgrounds, where US and German kiddies show democracy in action," boasted an article in the *HICOG Monthly Bulletin*.[54] American official rhetoric no longer equated Dachau with its concentration camp.

It is difficult to evaluate the immediate impact of the American confrontation policy. In response to Eisenhower's exhortation, "Let the world see," American and British media inundated their countries with photographs and film footage taken during the liberation of the German concentration camps. Atrocity propaganda corroborated the American and British wartime characterization of the Third Reich, revealing the extent of its criminality and rendering the Nazi regime synonymous with atrocity. In so doing, the photographic evidence of the camps served retroactively to justify the war and the occupation of Germany.

The response of the German civilian population to American atrocity propaganda was more complex. American intelligence found that the Germans accepted the photographs of the concentration camps as evidence of Nazi criminality but dissociated themselves from the crimes. After two years of occupation, the *Information Control Review* concluded that Germans still rejected the concept of collective guilt.[55] Worse, the Americans quickly realized, the confrontation policy was a political failure, alienating the Germans and making the Soviet military occupation look benign and sympathetic.

There is no doubt, however, that the atrocity policy succeeded in the long term. The U.S. Army photographers and cinematographers produced the first widely distributed images of the Holocaust, images that would acquire iconic status. Even now, the record they created remains one of the prevalent ways in which the Holocaust is visually represented. In the words of David Crew, "The way we remember Nazism is profoundly visual. When we try to imagine the Holocaust, it is frequently photographs—of piles of corpses, of skeletal survivors behind barbed wire—that come most rapidly to mind."[56]

Two

American Propaganda Films

During World War II, the Roosevelt administration called on Hollywood to produce movies to entertain, politicize, educate, and train both U.S. troops and those on the home front. The collaboration continued after the war. In American-occupied Germany, between 1945 and 1949, radio and movies were the main media of mass propaganda. Hollywood, the most "effective salesman for American products in foreign countries," was enlisted to help in the democratization project.[1] Later, with the emergence of the Cold War, Hollywood films became a tool of anti-Soviet propaganda, used to denounce Communism and proclaim the virtues of democratic capitalism.

Harold Zink, the official historian of OMGUS, asserts that "there was no organized course of indoctrination or propaganda" in OMGUS films.[2] The evidence, however, indicates that film played an integral role in the American propaganda campaign in Germany. The film program for Germany was controlled by the War Department's Civil Affairs Division (CAD) and Information Control Division (ICD). Political officers in CAD and ICD selected the films to be shown in the American zone and sector. Some were Hollywood movies or New Deal documentaries, and others were specifically produced by OMGUS for use in Germany. All were chosen to convey a new image of the United States and of Germany, and to transmit the American vision of the postwar world.

The evolution of the American film program reflects the existence of political and ideological tensions within and between OMGUS and CAD. OMGUS, in particular, was a heterogeneous organization that included military personnel and civilian employees with diverse political backgrounds. There were multiple views of the road to democratization, and different conceptions as to what image of America should be projected to the German public.

Documentary film as propaganda

In the era before television and before the Internet, film was the most important visual instrument of mass indoctrination. Its ability to intertwine fantasy and fact without losing the ring of authenticity made it an ideal medium for political propaganda.

Film was born as documentary. In 1895 Auguste and Louis Lumière produced the first of their short films, durable records of everyday events, and spectators were fascinated. Film allowed people to see things that had happened elsewhere, as if they themselves were there. From these short documents evolved newsreels, which developed alongside fiction film. The black-and-white documentary film in the 1930s and 1940s had a classical format—the narrative was bracketed between an introduction and a conclusion, and described a social or political problem and its solution.[3]

Nonfiction films were quickly recognized as an extraordinary tool for entertainment, education, and propaganda. Despite presumptions of objectivity and unmediated authenticity—in the 1930s, nonfiction films were called lecture films—documentary film is an interpretative device. To dramatize real-life events implies placing emphasis by deliberate design. The director selects and edits images, words, and music to advance a claim, to suit political objectives and aesthetic criteria. The editing reflects the ideology and influences of the society in which the film is produced and of the institutions—political, military, commercial, academic, religious—that financed, inspired or mandated its production. The manipulators of images and words cut, arrange, and splice the data, creating a harmonious montage.

The nonfiction film for domestic propaganda was a British political invention. In 1928 colonial secretary Leopold Amery, who chaired the Empire Marketing Board (EMB), launched a documentary film project to bring "the Empire alive to the imagination of the public." The Amery initiative, strongly encouraged by Rudyard Kipling, was developed by John Grierson. Grierson discovered the educational and propagandistic importance of the nonfiction film and became its prophet, its theorist, and its critic. For Grierson, a nonfiction film was "the creative interpretation of actuality." "Education," he stated, "has given [the public] facts but has not sufficiently given them faith." Nonfiction film could reduce that deficit.[4] EMB produced Grierson's *Industrial Britain,* a documentary that dramatized industrial work, and other films that exalted the contributions of miners, fishermen, and the postal service. During World War II, the Ministry of Information carried on the EMB tradition and produced an extensive series of short and long documentary films.

British nonfiction films had a strong impact on American political life. In 1935 the Roosevelt administration launched a propaganda initiative to justify its programs and publicize the successes of the New Deal. The Resettlement Administration (RA), created to help farm families relocate and assist them with loans, sponsored radio and photography campaigns and financed the work of artists and authors such as Walker Evans, Dorothea Lange, Ben Shahn, and James Agee. In 1934 Pare Lorentz—the key figure in the development of American nonfiction film—was appointed the RA film consultant. A master at blending aesthetics with politics, he produced two important films—*The Plow That Broke the Plains* (1936) and *The River* (1937)—for the U.S. Department of Agriculture. In 1938 Lorentz persuaded President Roosevelt to create the United States Film Service (USFS).[5]

Under Lorentz's direction, the new agency began to produce propaganda and educational films for commercial release. These films informed the American public about their rights and helped government employees tackle social problems. Lorentz directed *The Fight for Life* (1940) and produced two of the most important political movies of the period—*Power and the Land* (1940) and *The Land* (1942). Most American documentary films of the New Deal dealt with domestic social and political issues: the Great Depression, unemployment, the Dust Bowl, electrification and automation, the struggles facing migratory workers and the urban poor. A few, however, analyzed international themes such as the Spanish Civil War, the Sino-Japanese war, and the advent of authoritarian regimes.

USFS was not without its critics. USFS productions enjoyed resounding economic success, causing Hollywood to view the agency as a competitor. William Hays, the head of the Motion Picture Producers and Distributors of America and Hollywood's main censor, tried to interfere with Lorentz's movies. The industry's Production Code Administration (influenced by the Catholic Legion of Decency), demanded that American films be devoid of political messages that conflicted with their ideology.[6] Republicans and anti–New Deal Democrats in the U.S. Congress also attacked USFS, which they saw as a direct extension of the Roosevelt administration. They tried to cut its funding, alleging that Lorentz was using public funds to make pro-Roosevelt propaganda, and in 1940 eliminated the agency.

The anti-Roosevelt forces also recognized the importance of nonfiction film in political propaganda. In February 1935 Henry R. Luce, the founder of *Time* and *Life* magazines—and until 1939 a staunch opponent of the New Deal and an apologist for Mussolini and his regime—launched *The March of Time,* a new kind of cinematic journalism that integrated newsreel, documentary, and dramatic representation. Each episode, shown in movie theaters, was an exercise in the reconstruction of reality, blending "real footage" with reenactments of superb technical quality. In Luce's words, the films were "a fakery in allegiance with the truth." *The March of Time,* like *Time* and *Life,* reflected Luce's ideology and political agenda. An estimated twenty million Americans and fifteen million international spectators saw the programs every month in the late 1930s and early 1940s.[7]

Nonfiction films in World War II

In World War II, nonfiction film became an important medium for propaganda, used by all the belligerents to educate and energize the domestic front, to train soldiers, and to expand their influence in neutral countries. Producers and directors were commissioned to prepare films that would convince, mobilize, and galvanize the public behind the cause of war. The quality of documentary films was measured by their persuasive power.

In the United States, the eighty-five to ninety million Americans who went to the movie theaters each week to see feature films were a ready audience. Film became the fundamental media to instruct the American people about the war and the nature of the enemy, and to construct the image of a resolute and united American people.

By 1940 there were many government propaganda outfits involved in manufacturing an image of America for the domestic and international public. The main groups were the Office of Government Reports, the Office of the Coordinator of Inter-American Affairs, the Division of Information, Fighters for Freedom, the Office of Civilian Defense, the Office of Facts and Figures, and the Office of Coordination of Information. In 1942 President Roosevelt merged the information-control agencies into the Office of War Information (OWI), under the direction of Elmer Davis. The OWI joined the Office of Censorship and the Production Code Administration in prescreening movies produced by Hollywood and censoring and shaping their content. In 1942, the *Government Information Manual for the Motion Picture Industry,* produced by OWI, gave Hollywood precise political directives to ensure that its feature films mobilized public opinion in favor of the American war effort. The Roosevelt administration asked Hollywood to project the image of a strong America, while simultaneously depicting the heroism of Britain, the depravity of the Hitler dictatorship, and the brutality of the Japanese imperialists.[8]

Nonfiction films—both military training films and films for civilian education—were also seen as essential to the American war effort. Frank Capra and his collaborators—Theodore Seuss Geisel (better known as Dr. Seuss), Joris Ivens, Anatole Litvak, Ernst Lubitsch, Dimitri Tiomkin, and Anthony Veiller—made a series of documentary films explaining why the United States had entered the war, the nature of the enemy, and what was at stake in the fight against the Axis.[9] Capra's *Why We Fight* series (*Prelude to War, The Nazis Strike, Divide and Conquer, Battle of Britain, Battle of Russia, Battle of China,* and *War Comes to America*) was produced by the Army Pictorial Service and shown to all troops as part of basic training. The films were later released nationwide and shown to the general public. Still classics of political propaganda, these films used animation, newsreel material, voice-overs, dramatic music, and captured enemy footage.

By 1943 the War Department had a full-fledged film program to teach soldiers about the causes and the importance of the war. Training films were meant to educate troops in military doctrine, teach them about the use and care of weapons, and instruct them on tactics, military law, and hygiene and sanitation. Lively productions that incorporated the most modern techniques of animation, popular music, and GI jargon replaced lectures, and proved to be an effective way to motivate the troops and to induce the will to fight.[10] In 1948, Emanuel Cohen, the executive producer in charge of Army films, explained the importance of cinema in the modern soldier's training program:

The generation drafted [in World War II], unlike their fathers in 1917, had grown up with motion pictures. Films were as much a part of their lives as books and bicycles and malted milkshakes.... Films, consequently, were a natural and familiar form of entertainment.... It was the ideal way of giving a message in familiar form and having it accepted.[11]

The film program cut training time by 30 percent because soldiers learned faster through film than through classroom work. General George C. Marshall, U.S. Army chief of staff, considered that the airplane and the motion picture were the two most important weapons developed during World War II.[12]

Films created for the home front were also key. "Incentive films," made by the War Production Board, stressed the importance of the American working class in the "Arsenal of Democracy." "Information films" such as *World at War, Geography of Japan, Colleges at War,* and *Doctors at War* instructed specific sectors of society in proper wartime behavior. A particularly distressing group of information films—among them, *Japanese Relocation* and *Challenge to Democracy*—sought to justify the racist repression of Japanese Americans.

Meanwhile, the overseas branch of OWI produced nonfiction film for foreign audiences. The unifying characteristic of these films is that they presented a superficial and sentimental perspective of life in America with an emphasis on small-town values, religiosity, and social harmony. Some offered a mechanical description of American politics and American institutions or provided vistas of "daily life" in the United States.[13] In most cases, documentary films made for export ignored the political problems besetting the United States at the time, making no mention of racism, discrimination, anti-Semitism, segregation, the corruption of the political class, the huge concentration of industrial and financial capital, gangsterism, and rabid anti-intellectualism.

American film policy in occupied Germany

The Nazis had also used film intensively for propaganda purposes. Indeed, John Grierson considered Hitler to be the master of modern propaganda, someone who understood that propaganda was the "very first and most vital weapon in political management and military achievement."[14] In March 1933 the Nazi regime set up the Ministerium für Volksaufklärung und Propaganda (Ministry for Popular Enlightenment and Propaganda) under the direction of Joseph Goebbels. The German film industry, which had been Hollywood's most important competitor in the prewar international market, was consolidated into a state monopoly, Ufa (Universum-Film-Aktiengesellschaft). In August 1940 the German government banned all American films in areas under its control, and Ufa became the sole source of films in the Third Reich and Nazi-occupied Europe.[15] The Nazis used film as an instrument of mass education and as a means of mass entertainment with great success. Attendance at movie theaters in Germany quadrupled during the Nazi period, growing from 250 million in 1933 to a billion in 1942.[16]

Nazi films constructed a parallel visual reality devoid of conflict. Only about one-sixth of the 1,097 feature films produced between 1933 and 1945 were overtly political *Tendenzfilme,* but all corresponded to the *Völkische*

Weltanschauung the regime was trying to create. Ufa productions ranged from light escapist comedies and musicals to virulent anti-Semitic propaganda films and "euthanasia" films.[17] Even so, Siegfried Kracauer has argued, "All Nazi films were more or less propaganda films—even the mere entertainment pictures which seem to be remote from politics."[18] All films, fiction or nonfiction, were laced with ideological subtexts. Ufa films depicted German citizens as racially homogeneous, hard working, and loyal, always ready for supreme sacrifice in the service of the *Führer* and the Fatherland. In late 1938, simultaneous with *Kristallnacht*, German film directors were given explicit orders to produce anti-Semitic films. In 1940 *Die Rothschilds, Jud Süss*, and *Der ewige Jude*, three major anti-Semitic propaganda films, were released.[19]

Nazi films did not show the concentration and extermination camp system, but they underlined the achievements of the Wehrmacht, exalting the modernity of German military technology and the strength of the German soldier. Every Wehrmacht regiment had a Propaganda Kompanie that photographed and filmed its participation in the war, providing Ufa with stills and footage from the front.[20] The Third Reich also used film in its effort to win over the population of neutral countries, especially in Latin America, to the Nazi cause.

As soon as the war ended, U.S. occupation forces closed all the German cinemas that survived the Anglo-American strategic bombing and ground combat, prohibited the showing of German films, and impounded all films made during the Third Reich. Many of the movies sequestered by the Americans were sent to OWI headquarters in New York, where they were stored as poisonous propaganda. Once OMGUS was established, it proceeded to break up Ufa.[21]

The Soviets and the British were the first to realize that it was necessary to provide entertainment to the German population in the dismal living conditions of 1945. The Soviet military administration rushed to clear away the rubble blocking access to theaters in Berlin and, in early July 1945, began to screen Soviet and German films. An American intelligence report from July 12 determined that "USSR officials believed that entertainment activities were necessary, at all costs. The people of Berlin needed to forget their hungry stomachs, their former homes . . . the whole gory picture of a defeated city."[22] *Variety*, the journal of the American entertainment industry, warned that "American film companies are being left at the post in Germany with Russia racing down a straightaway field in utilizing pictures as a propaganda medium for their political philosophy."[23] The British also began to reopen movie theaters in their zone and sector in early July 1945. Gladwin Hill, an American war correspondent and film critic, suggested that OMGUS was actually surprised by the British decision:

> The SHAEF Information Control section, a joint Anglo-American organization, was originally planned to suppress all ordinary motion picture exhibition in Germany for a long period and gradually to introduce a new era of cleansed, non-Nazi films of both foreign and German origin. But within two months after V-E Day the German people were getting so restless, virtually confined to their homes, that Marshal Montgomery became worried about keeping order in the British zone. He abruptly pulled out of the joint program—with an embarrassing lack of notice to the American section—and reopened the German theaters with any available pictures that seemed harmless.[24]

As often happened in cultural matters, the Americans were the last to react, but by the end of July, they too had begun to reopen movie theaters. By the end of 1945 ICD had opened 350 movie theaters in the American zone and sector, and more that 500 more theaters were scheduled to open in 1946. In Frankfurt, ICD officers discovered more than eight thousand reels of German feature film in vaults, but very few of them were considered acceptable for screening. Therefore, OMGUS made newsreels for the German public and brought in material from the United States. ICD used the large UFI (Universum Film G.m.b.H.) studios in Munich and Berlin to produce the Anglo-American weekly newsreel *Welt im Film* ("World in Film"). OMGUS also showed a selection of American newsreels, documentaries, and Hollywood feature films.[25] German productions were eventually incorporated into the film program.

By September 1945 thirty Hollywood feature films (between two and five years old)—the "invasion films"— had been included in the reeducation program for postwar Germany. Hollywood had much to gain from continuing its partnership with the government in the postwar period. Prewar Europe had been a very important market for Hollywood films, but American films had been barred from Nazi Germany and Nazi-occupied Europe in 1940. Postwar Europe provided an immense opportunity for the industry, and Hollywood was eager to position itself as strongly as possible in the European market.[26]

The planners of the U.S. occupation strategy, like their Soviet counterparts, viewed film as a means of political reeducation, and not just as a means of diversion.[27] Entertainment was a secondary concern. Brigadier General McClure, the head of ICD, stated that the OMGUS film program's long-range objective was to "re-orient the German mind after 12 years of Nazism." Robert Joseph, who served as OMGUS film officer for Berlin and as deputy film officer for Germany until 1947, was even more explicit: "If the Germans are entertained by what they see, this is merely an unimportant adjunct of the over-all film program for the German Reich."[28]

ICD used Hollywood films to establish a picture-perfect America for German consumption. Films were prescreened by the film censor at the Office of the Director of ICD, which, in conjunction with the State Department and the War Department, evaluated their strategic usefulness. These agencies selected Hollywood films of a "positive nature" that would "protect American interests and maintain American prestige abroad."[29] The films were intended to challenge Nazi stereotypes about America and promulgate messages that would "counteract the teachings and doctrines of National Socialism [and] enlighten the German people as to what went on in the world outside Germany during the last twelve years."[30]

Feature films determined to be "of questionable moral contents" were not shown in the American zone and sector, nor were "horse opera type films, westerns . . . and films dealing with gangsterism, crime."[31] Films depicting slavery were banned for perpetuating negative images of American society; this was the case with *Gone with the Wind.* The celebrated film adaptation of Steinbeck's *The Grapes of Wrath* was deemed subversive, as was *The Right to Strike. The Maltese Falcon,* the first American film noir, directed by John Huston and starring Humphrey Bogart, was rapidly withdrawn; ICD censors feared it would be read by the German public as a glorification of the criminal world and a critique of the American police.[32]

Documentary films were also an essential part of the OMGUS film program. Films such as *Tennessee Valley Authority, Autobiography of a Jeep, The Town,* and *Democracy in Action,* all produced for American audiences in the 1940s, were translated into German and shown in movie theaters, American information centers, schools, and town meetings.[33] These films highlighted the virtues of grassroots democracy and social cooperation, introduced civic and egalitarian values, and stressed the virtues of tolerance, cooperation, and solidarity. In 1951 Henry P. Pilgert reflected on the political role of American documentary film in occupied Germany:

Documentary evidence of life . . . in the United States, through the media of films, reveals the construction and accomplishments of democratic governments and their peoples, their geographical and economic situation, their attitudes, their way of living, their cultural, professional, educational, industrial and technical developments, as well as important past and present events.[34]

Like their fiction counterparts, the American nonfiction films shown in Germany presented an idealized image of the United States as a just, ordered, god-fearing, and harmonious society of abundance and fairness.

ICD considered the documentary film program a political and educational success, inferring a causal connection between the new German civic attitude and the OMGUS screenings. Writing in HICOG's *Monthly Information Bulletin,* Haynes R. Mahoney claimed that American film had had a tremendous impact in Germany:

> Never before in German history has the average man from "Main Street" taken such an interest and wielded such influence in the activities of his government—at least on local levels. And who can say how many ideas may have been planted in the minds of simple burghers, of parents and teachers and youth by "TVA," "County Agent," "The Story that Couldn't be Printed," "Lessons in Living," "Playtown, U.S.A," or many others of the broad variety of films circulating throughout the U.S. Zone of Germany.[35]

Such enthusiastic reports may, however, simply reflect the self-congratulatory style of the bureaucratic machine. Not all American film experts agreed with the military's appraisal of its film program. Gladwin Hill, a war correspondent and film critic for the *New York Times,* argued that the OMGUS film program was vacuous:

> Documentaries . . . had not stirred a ripple among the German people [and the] Anglo-American newsreels . . . made American bathing-beauty-and-dogshow nonsense look like Academy winners by comparison.[36]

Paul Linebarger, an expert on psychological warfare, gave an equally dismal appraisal:

> Few propaganda movies have ever achieved the spectacular impacts of some private films in portraying the American way of life. Tahitians,

Kansu men, Hindus and Portuguese would probably agree unanimously in preferring the USA of Laurel and Hardy to the USA of strong-faced men building dams and teaching better chicken-raising.[37]

Todesmühlen, an atrocity film for Germany

The first OMGUS documentary made specifically for Germany was *Todesmühlen* ("Mills of Death"), an atrocity film incorporating footage taken by the U.S. Army Signal Corps and Soviet camera units during the liberation of Nazi concentration and extermination camps in Germany and Eastern Europe. The complicated history of *Todesmühlen* shows the rapid shift in American policy in Germany. Initially, the film was conceived as a central element of the confrontation policy, but after just a few months ICD lost interest in the project. By early 1946, when *Todesmühlen* was released, OMGUS had already moved away from the hypothesis of German collective guilt.

After the discovery of the concentration camps in Germany, SHAEF's Psychological Warfare Division instructed the photographers and cameramen of the Signal Corps to visually document what they encountered. They were asked to show graves and to record the grisly piles of objects such as garments, shoes, jewelry, and false teeth.[38] Thus, the most shocking elements of the German concentration camp system—emaciated corpses, mass graves, crematoriums, gas "showers," surviving inmates in advanced degrees of starvation—were registered for posterity. The Signal Corps also recorded the aftermath of liberation—the mass burials and the confrontation policy. Although the footage of the camps was taken in a period of extreme confusion, the amount of visual information gathered was unprecedented in the history of wartime filmmaking. This material was used in war crimes trials, in newsreels, in shorts such as *KZ*, made for the founding conference of the UN in San Francisco, and in *Todesmühlen*.[39]

Immediately after the discovery of the concentration camps in Germany, Elmer Davis, the OWI chief in Washington, announced that American propaganda would "hammer to the German people the atrocities being disclosed."[40] *Todesmühlen*, tailored for the German public, fit this agenda—the film was expected to induce a sense of collective guilt in German viewers. Politically, it was meant to provoke a renunciation of National Socialism and make the Germans accept defeat and occupation as deserved punishment for abetting the criminal activities of the Third

Reich. This, in principle, would dissuade Germans from organizing and joining terrorist or guerrilla movements fighting the Allied occupation.

The production of *Todesmühlen* was fraught with conflict. Initially, the film was planned as a coproduction with the British Ministry of Information (MOI), to be directed by Sidney L. Bernstein, the London head of SHAEF's PWD (Rear) Film Section. Bernstein had trouble gaining access to the Signal Corp footage, however, because American intelligence officers were noncooperative. He concluded that production was being intentionally obstructed. In fact, Davidson Taylor, head of PWD Film, Theater and Music Control Branch in Berlin, believed that the film would be a strategic mistake and would not succeed in convincing the German people of their collective guilt. Taylor asked Brigadier General Robert A. McClure, chief of PWD, to remove Bernstein and MOI from the *Todesmühlen* project and proposed that PWD take over production. General McClure approved the recommendation, and PWD took exclusive charge of the atrocity film.[41]

Still, the production was hindered by disagreements concerning the concept and the script. In the summer of 1945 Hanus Burger, a member of PWD attached to the 12th Army Group, was sent to London to select footage for the film. Burger was born in Prague to German Jewish parents, but he had studied in Frankfurt and been involved in theater in Hamburg. In 1931, in view of the worsening political situation in Germany, he returned to Prague, where he worked in theater. He was a Communist, and on occasion served as a courier for the Czech Communist Party. While in Prague he met the American producer and director Herbert Kline, and together they began to work on *Crisis,* a documentary about the destruction of Czechoslovakia after the *Anschluss,* the March 1938 Nazi-Austrian putsch. The film was never finished because of the increasingly dangerous political situation, and in late 1938 Burger was forced to leave Prague for France. There, he was able to edit *Crisis* before leaving for the United States. He settled in New York, where he directed the short documentaries *Portrait of a Library* (1940), *It Happened Here* (1941), and *Education for Tomorrow* (1941), as well as *Seeds of Freedom* (1943), a remake of a Soviet film produced by Sergei Eisenstein in 1925. In 1941 Burger became a U.S. citizen and joined the U.S. Army, but he was recruited by the OSS, which was very interested in German-speaking exiles. Before D-Day he was transferred to PWD/SHAEF, and was part of a mobile broadcasting unit that participated in the Normandy landing. Later, he was assigned to Radio Luxembourg and to Radio 1212—a radio station run by the OSS that aired black propaganda.[42]

Burger saw the *Todesmühlen* project as an opportunity to confront German viewers with Nazi atrocities and make it difficult to deny the existence of the camps. His intention was to give an account of German fascism from 1931 to 1945, highlighting both the evolution of the Nazi criminal project and the participation of the German population. Moreover, he wanted to document the role of capitalism and heavy industry in the rise and development of Nazism, as well as the link between the concentration camp system and German industry.[43]

PWD asked Billy Wilder to review Burger's feature-length edit. For a few weeks in August 1945, Wilder served as an OMGUS film officer in Berlin, acting as an expert consultant for the denazification and reorganization of the German film industry.[44] Burger respected Wilder as a great director, but wondered if a man who had spent the "difficult" years in California, and who had shown no compassion for the victims of Nazi brutality in any of his films, was really the best person to finish a project like *Todesmühlen*. Burger tried to dissuade his superiors, emphasizing that Wilder had not participated in the war effort.[45]

Burger's reservations proved justified. Years later, Burger wrote that Wilder did not like the film. He was not particularly impressed by the concentration camp images—the inmates, he said, appeared to be wearing pajamas—and wanted to shorten the camp footage to only four seconds. In Wilder's opinion, the German people were sick and tired of the past, and the Americans should stop beating them over the head with films like *Todesmühlen* and instead focus on producing films that would entertain in "intelligent ways." Wilder demanded that Burger's two-hour film be cut and reedited in a week. A twenty-two-minute version was completed by October 1945.[46] Although the film no longer emphasized the role of German capital in the Third Reich, the notion of popular complicity in Nazi crimes remained strong.

The tensions between Burger and Wilder were likely personal as well as political. Both were European Jews who had fled to the United States, but their professional and existential experiences, and their political ideologies, were radically different. Burger was a Communist, and Wilder was not. This helps explain their conflicting interpretations of the role of capital in the emergence and consolidation of Nazism. Moreover, Wilder's experience in the United States had been very different from Burger's. Wilder was born into a Jewish family in Sucha, then part of the Austro-Hungarian Empire. He had studied law in Vienna but quit the university to become a journalist in Berlin; his cinematographic career began with his collaboration in the film *Menschen Am Sontag* (1929). Wilder left Germany in 1933, immediately after the Reichstag fire, and settled in the United States. Although he arrived penniless and without knowing English, he made a conscious effort to learn the language and embraced American culture with a passion. He took Hollywood by storm, creating a sensation as a director, producer, and screenwriter. In 1941 he directed *Hold Back the Dawn*, and in 1944 *Double Indemnity*. By the time Wilder returned to Berlin with OMGUS, he was regarded as one of the greatest directors in Hollywood.[47]

Yet the controversy around *Todesmühlen* allows for a much richer, deeper, and more disturbing reading of the clash between the two directors. Billy Wilder's mother, stepfather, and grandmother had been murdered in Auschwitz. This fact is well known—it was repeated in all his obituaries—yet scholarly accounts of *Todesmühlen* do not even mention it in their analysis of the film. Wilder himself was curiously detached about the fate of his family. He later recalled that he learned of their fate in Berlin in 1945. "I don't know anything about how she [Wilder's mother] ended up in the camp," he told a journalist. "I just know that it was Auschwitz because everybody from Vienna, where she lived, went there. I did not know that they had concentration camps. You know, it was kept quiet. Roosevelt did not tell us about it."[48]

Wilder's seeming indifference to the Nazi atrocities typifies the reaction of many European exiles. Wilder distanced himself from his European past and from the Holocaust. He directed over fifty films during his career, worked with the best actors and actresses of the era, and received six Academy Awards, dominating Hollywood from the late 1930s to the early 1960s. Yet none of Wilder's films even touched upon the Holocaust. His postwar films that dealt with Europe, like *A Foreign Affair* (1948), do not hint at the Nazi crimes against the European Jews. While this is not unique—there were no major Hollywood films that dealt with the Holocaust until the 1960s—it is worth noting given his personal history. Wilder seemed to cling to the vision of a good Germany, victimized by Nazism, that would emerge untainted from the war.[49] As Marlene Dietrich, playing Erika von Schlütow in Wilder's *A Foreign Affair*, sings: "A brand new spring is to begin, out of the ruins of Berlin."

Todesmühlen is an ordinary documentary film dealing with an extraordinary phenomenon. It shows images from Auschwitz, Sachsenhausen, Maidanek, Belsen, Ohrdruf, Mauthausen, Ebensee, Ravensbrück, Dachau, and Buchenwald but makes it clear that these camps were just a few of the three hundred German death mills scattered around Europe. The film includes images that would become iconic: the sign reading *Arbeit macht*

Frei, the living skeletons, the mountains of corpses, the mass graves, the barracks, the crematoria, the gas chambers, the Zyklon B containers. The viewer witnesses a succession of piles: piles of corpses, of bones, of clothing, of shoes, of children's toys, of hair, of gold, of jewelry, of earrings, of eye glasses. General Eisenhower is shown visiting a camp, a figure of authority and justice contemplating the remains of the German killing machine and its victims.

The film does not comment on the Nazi's anti-Semitism. The inmates, the film explains, came from all European nations, "Russians, Poles, Frenchmen, Belgians, Yugoslavians, Germans, Czechs" and belonged to "all religions: Protestants, Catholics, Jews. . . . They were murdered because of their religion or because they did not want to be Nazis, or because they were Russian, Polish, Belgian, Czech, or Jewish." The genocidal campaign against the European Jews is not highlighted, an omission that continued the American wartime policy of minimizing the Nazis' focus on exterminating the European Jews. It is evident that neither CAD nor ICD exerted pressure on the filmmakers to emphasize the role of Nazi anti-Semitism. Personally, Burger adhered to the Communist line that ignored the specificity of the Jewish Holocaust, and Wilder, like so many assimilated Jews, was not particularly interested in the fate of the European Jews.[50]

The film is less vague in its discussion of the perpetrators. There were two kinds of criminals, it suggests: the SS personnel directly involved in the atrocities, and the German civilians who tolerated them. The German people are collectively indicted, accused of having enthusiastically and fanatically supported Hitler and of having acquiesced to the mass torture and extermination of innocent people in the name of their country.

Todesmühlen—the shortened version—was shown in November 1945 to an ICD audience in Bad Homburg and tested with a German audience in Frankfurt. Then OMGUS lost interest in the film. After these preliminary screenings, the production credit was changed from "Office of War Information" to "Allied Army Cameramen," implying a distance between the American occupiers and the film. *Todesmühlen* was shown in Bavaria in January 1946, throughout Greater Hesse, Baden-Württemberg, and Bremen in March 1946, and in Berlin in May 1946. The openings were was not preceded by a significant publicity campaign. The film ran in each theater in which it was shown for only one week, as part of a sixty-minute program that included the current issue of *Welt im Film* and another American documentary. Attendance was strictly voluntary, children under age fourteen were not admitted, and spectators were charged normal newsreel admission prices.⁵¹ In fact, when OMGUS officers in Bavaria made viewing *Todesmühlen* compulsory in their area, ICD reiterated the reminder that "such practice is contrary to Information Control Division's policy and the overall policy of Military Government."⁵²

The film was not popular with German viewers. Monitoring audience reaction, American intelligence found that Germans were significantly less interested in *Todesmühlen* than in other American films.⁵³ A drawing by Gerhard Kreische in the satirical journal *Ulenspiegel* underscores the alienating effect of the film. It shows two parallel scenes—a crowd queuing outside a renovated movie theater showing Leni Riefenstahl's *Der Sieg des Glaubens* ("The Victory of Faith"), and an empty, rundown movie theater showing *Todesmühlen*. A young woman walking a dog strolls by the *Todesmühlen* poster without even turning to look.⁵⁴ In Munich, where even old movies tended to sell out, the turnout for *Todesmühlen* was extremely low—barely one-third of the seats were filled. In the American sector of Berlin, fewer than half of the seats were sold on opening day, and the audience dropped drastically in the days that followed. The average weekly movie attendance in the American sector of Berlin was 250,000 people (26 percent of the total population), but during the week devoted to *Todesmühlen* attendance fell to 157,120 (16 percent).⁵⁵ OMGUS polls indicated that the majority of German viewers were not persuaded by the film's emphasis on the notion of collective guilt. In Berlin, 70 percent of the 1,040 respondents continued to believe that the German people did not share responsibility for what had happened even after seeing the film. However, 82 percent of respondents claimed to have learned about the existence of death camps through *Todesmühlen*.⁵⁶

Judged by psychological warfare criteria, the film violated the main tenet of advertising—the propagandist must not antagonize the target audience. *Todesmühlen* clearly antagonized viewers and made them react against its message. Moreover, it was perceived as propaganda, so its veracity was automatically suspect. Michael Josselson, an OMGUS intelligence officer at the time, considered overt propaganda films counterproductive in Germany:

> The German public has been fully saturated with propaganda films during the last 12 years and therefore is very skeptical of any kind of propaganda today. This is best evidenced by the unpopularity of Soviet films shown in Berlin. . . . The conclusion may therefore be drawn that showing a series of straight propaganda films in Germany at this time will not achieve the desired effect.⁵⁷

An OMGUS *Weekly Information Bulletin* from late February 1946 concluded that *Todesmühlen* "was decidedly not a box-office success."[58]

Yet *Todesmühlen* did succeed in bringing the camps into the German public visual sphere. This challenged the Hitler/Himmler project of clandestine extermination and established the reality of Nazi crimes against humanity. Individual Germans could condemn or justify what they saw on screen, but showing Nazi atrocities transformed a foggy allegation of misconduct into a tragic reality. ICD was aware of the imporatnce of *Todesmühlen,* as indicated in a January 1946 *Monthly Report:*

The film constitutes one of the most damning indictments of the Nazi system to reach the public. Audiences at test showings involuntarily gasped and groaned when certain scenes came on the screen, and many who had never before believed tales of Nazi atrocities confessed themselves to be convinced by the incontrovertible testimony of the camera.[59]

Another issue of the OMGUS *Weekly Information Bulletin,* from December 1946, claimed that *Todesmühlen* "had a decided impact on German audiences."[60]

The ICD Documentary Film Unit

By the time *Todesmühlen* was released, OMGUS was also worried about the phantom of civil unrest and aware of the threat posed by Soviet propaganda. The Soviet military government, from the very beginning of the occupation, had skirted the issue of German collective responsibility. The emergence of the Cold War brought a strategic shift in OMGUS propaganda, which was reflected in its film policy. The American propaganda message became lighter in tone, building on the idea of German reconstruction and the consolidation of German-American relations.[61] After *Todesmühlen,* the goal was to produce movies that would gain the hearts and minds of the German population.

In August 1947 ICD formed the Documentary Film Unit (DFU), a propaganda unit charged with making nonfiction films for the German public that addressed American Cold War interests. The unit was led by Stuart Schulberg, a young Marine Corps sergeant who had directed top secret training films for the OSS. In 1946 Schulberg joined a film unit headed by John Ford. He and his brother Budd were sent to Berlin with the mission of gathering film and photos for use in the Nuremberg Trials. Stuart Schulberg remained in Berlin until 1949, when he left for Paris to direct the Marshall Plan Economic Cooperation Administration Film Unit. Between 1947 and 1949 DFU produced more than a dozen films.[62]

According to Stuart Schulberg, the film program's primary concern was the "political, social, and economic reorientation of the German people."[63] Lieutenant Colonel Pare Lorentz, chief of Films, Theater and Music in the War Department's Civil Affairs Division from 1946 to 1947, collaborated with DFU.[64] DFU films fall into two distinct categories—anti-Nazi films and Cold War propaganda films. The anti-Nazi films stressed the causal connection between German aggression and postwar European misery and insisted on the complicity of the German population with the Third Reich. The Cold War films did not antagonize Germans with references to Nazism but rather emphasized American-German cooperation, explaining the Blockade, the Berlin Airlift, and the Marshall Plan from the American anti-Soviet perspective. The *Information Control Review* for December 1947 explained that "Military Government has intensified its political information program in order to reemphasize and reaffirm the principles of democracy and to warn the German people of Communistic dangers."[65]

Schulberg's *Hunger* (1948) is a prototypical DFU anti-Nazi film. Like *Todesmühlen,* it broke the cardinal rule of advertising by confronting its target audience with unpleasant information. *Hunger* shows that Germans were not the only ones suffering in the postwar period. It opens with images of children scavenging for food as a narrator explains, "No, this is not Stuttgart—it's Lodz, Poland. And this is not Munich—it's Naples, Italy." The litany continues as the image changes to scenes of emaciated women searching for food among the ruins: "This is not the Ruhr—it's Greece. This is not Berlin—it's Paris." This went against the rhetoric of German particularity and singular victimhood that emerged in the immediate postwar period in Germany.[66] *Hunger* places the blame for European and German devastation on the Nazi regime, and on the Germans who stood by it. The film warns that the recovery process will be slow and difficult. Yet it does not discuss the Holocaust even in passing, even though the concentration and exterminations camps were the paradigmatic example of starvation in Europe.

Hunger drew complaints from its German audiences. There was whistling in the theaters, newspapers published letters to the editors complaining about the film, and OMGUS offices received phone calls expressing dissatisfaction. Schulberg recalled some typical comments:

You can't tell us anyone else is as hungry as we are and you shouldn't try.

If you spent more time bringing in food and less time making propaganda films, we'd all be better off.

The war has nothing to do with hunger. We ate wonderfully during the war. Germany is being starved because we haven't the money to pay your food prices.

The film is phony. The reason we have no food is simple: You gave our best agricultural land to Poland.

According to Schulberg, "almost all blamed the crisis on inefficiency, Morgenthauism, greed or downright cruelty."[67] Intelligence surveys indicated that many Germans believed the Communist claim that the United States was withholding or destroying food in order to drive food prices up.

The most disturbing response to *Hunger* was the enthusiasm with which some in the German audience reacted when they saw images of Nazi leaders and the German military machine. A scene showing German troops marching to war was heartily applauded throughout the zone, specially by young people, even though it was accompanied by a satirical sound track. An image of Hermann Göring elicited different reactions in different parts of Germany. In West Berlin many laughed, in Munich the audience kept a sepulchral silence, and in small towns in Bavaria and Baden-Württemberg there were cries such as "Good old Hermann," "Our Hermann!," "We need a new Hermann—he wouldn't let us starve."[68]

ICD had planned to show *Hunger* for twelve weeks, but OMGUS withdrew the movie from theaters in its third week. According to Schulberg, "The official excuse was that the entire food information campaign had been 'overdone,'" and the order to withdraw the film came directly from General Clay. Schulberg, however, believed that the film was pulled because it was "too anti-Nazi," and because "it didn't fit into the new campaign of blaming everything on Russia and the SED [the German Socialist Unity Party]." [69]

By contrast, *Die Brücke* ("The Bridge," 1949), a DFU Cold War film, uses the Berlin Blockade to emphasize the growing collaboration between Western Germans and Americans. The massive airlift is shown as heroic, effective, and politically meaningful because it consolidated this postwar bond. The heroes are the American and British pilots, but the film does more than highlight the immense technological and logistic power of the United States. The narrative centers on the personal relationship that develops between an American pilot and a German who works at Tempelhof unloading planes. At first the two men belong to different worlds and only meet on the tarmac. By the end, however, they have grown to trust and to respect one another. The narrator concludes, "That's the story of the air bridge. More than food and coal are coming in. Friendship is coming in too. And belief. Belief in tomorrow. Sometimes I think this is not an air bridge at all. It is a human bridge, linking together people who want to be free. A brand new kind of understanding."

The film *Zwischen West und Ost* ("Between West and East," 1949), a rhapsody to postwar West Berlin and its people (to whom it is dedicated), is particularly interesting as an example of the extensive rewriting of history through nonfiction film. *Zwischen West und Ost* offers a narrative of the political events that shaped 1945–1948 Berlin but minimizes the role played by the United States in destroying the city and defeating the Third Reich. The initial shots show Berlin from the air—a landscape of ruins and cranes. The narrator attributes the reconstruction of the city to the hard work of the 2.25 million West Berliners determined to turn the capital of Nazism into an outpost of democracy. The origins of this process are traced to 1945, when the "people" rebelled against the "crooked philosophy put into their minds for 12 straight years." West Berliners had had "enough of force, enough of terror" and did not want a new dictatorship, either "brown or red." The film describes denazification as a German initiative. West Berliners, the narrator claims, "had destroyed the swastika, and now they were determined that the hammer and the sickle would never stand in its place." Germans thus emerge as heroes of the resistance against Nazism and as staunch defenders of liberal democracy. The Americans appear as the protectors of German democracy and the willing partners of the West Berliners fighting for their freedom.

Zwei Städte ("Two Cities," 1949), the last DFU film produced for Germany, does not even mention Nazism, focusing exclusively on anti-Soviet propaganda. The film compares the condition of two German cities, Dresden (Soviet zone) and Stuttgart (American zone), four years after the end of the war. Dresden, destroyed and unreconstructed, is depicted as "the story of yesterday," Stuttgart, thriving and reconstructed, as "the story of tomorrow." Dresden is covered with political posters, but there is no food in the stores. In Stuttgart there are few political posters but plenty of food. Dresden's stagnation is attributed to Soviet ideology and ineptitude, while Stuttgart's recovery is an "accomplishment of freedom" made possible by economic liberalism, a solid currency, the participation of Western Germany in world trade, Anglo-American aid, and the Marshall Plan.

The Brotherhood of Man

Yet the construction of a positive image of the United States for German consumption was not easy because the question of *which* image of America to export was highly contested at home. World War II was the turning point of American military and economic power, and the political establishment was interested in projecting American culture abroad. The Roosevelt and Truman administrations had a decidedly anticolonialist rhetoric, emphasized the need to promote democracy abroad, and exalted the values of American political and economic liberalism. Yet institutional racism in the United States threatened the credibility of the American message abroad. Jim Crow continued in the American South, and segregation and discrimination still plagued American society. The number of lynchings diminished in the postwar years, but racial violence and ghettoization persisted in American cities.

Racial tensions also existed within OMGUS. In Germany, racial conflicts within the occupying forces surged after V-E Day. The Army tended to attribute violations of discipline to the alleged inferiority of African American soldiers, and reports to its Criminal Investigation Division and to the Senate's War Investigation Committee both advised their rapid recall from Europe. At the same time, the War Department was under pressure by the White House to desegregate the Army and improve the condition of black soldiers stationed abroad. In May 1946 a delegation from the Negro Newspaper Publishers Association visited Germany and reported that there were no African Americans either in the American military government or in USFET. General Clay believed that black units should be restricted to parades; General Ernest Harmon, the commander of the constabulary, only accepted white soldiers for his police unit; and General Keyes, the successor of General Patton as commander of the Third Army, tried to reduce the number of African Americans among his troops. Rapid demobilization, however, gave priority to soldiers who had been in combat; this raised the proportion of black soldiers in Europe, because most of them had been excluded from combat duty. Moreover, African American soldiers, not eager to face antiblack violence at home, reenlisted at a higher rate than white soldiers. By 1947, black soldiers comprised 15 percent of the U.S. Army (the wartime high had been 9.68 percent). When the War Department ordered an increase in the proportion of African American soldiers in the European Theater, a military delegation from Europe went to Washington to propose the reduction of the assigned percentage (from 15 percent to 8 percent). After negotiations, it was agreed that black soldiers would be assigned exclusively to subordinate tasks, and never deployed in infantry units, in the constabulary, in highly technical services, or as supervisors of white civilians. In Germany, General Keyes ordered the exclusion of African American personnel from administrative positions—including even the postal service—at group level in the Third Army.[70]

With the end of the antifascist alliance and the emergence of the Cold War, the race issue reappeared in Soviet anti-American propaganda, and in Communist propaganda all over the world. The U.S. State Department estimated that 50 percent of Soviet anti-American propaganda in the immediate postwar period focused on American racism.[71]

Racial tensions also plagued the American industrial workforce. In 1946 Walter P. Reuther, president of the UAW-CIO, produced *The Brotherhood of Man,* an animated antiracist cartoon, as part of a political education campaign to lessen racial animosity in the American labor movement.[72] Pare Lorentz, chief of the Films and Theater Section of CAD's Reorientation Branch, recommended that the film be shown in occupied Germany, as a "light treatment of serious themes of racial tolerance" that would fit within the OMGUS reeducation program. In August 1946 CAD New York bought more than two hundred copies of *The Brotherhood of Man* and translated it into German.[73] However, the War Department vetoed the screenings in 1947, and the film was never shown in Germany.

The Brotherhood of Man was based on Ruth Benedict and Gene Weltfish's 1943 pamphlet *The Races of Mankind.* Benedict had been a student of Franz Boas, a German-Jewish emigré who became a central figure in American anthropology and the leading advocate for open immigration policies.[74] During the war, Benedict was part of the Division of Cultural Analysis and Research of the Bureau of Intelligence, OWI, where she studied Japanese and German culture; Weltfish was in the Office of Strategic Services (OSS).

Written in plain language and illustrated with cartoons, *The Races of Mankind* provided an account of genetics that was accessible to a layperson. The pamphlet refuted the Nazi claim of Aryan superiority and showed that the Nazi pseudotheories of race had no scientific basis. The authors put special emphasis on refuting the notion of black intellectual inferiority, citing the results of Army intelligence tests that showed that the median score of northern blacks on intelligence tests were significantly higher than those of southern whites. Benedict and Weltfish concluded that differences in "intelligence" were not attributable to race, but rather to socialization.[75]

In January 1944 Lieutenant Colonel Arthur C. Farlow, the head of the

Orientation Branch of the U.S. Army, issued an order for fifty-five thousand copies of the pamphlet to be included in the orientation kit distributed to GIs.[76] The United Services Organizations (USO), a group of six civilian agencies created in 1941 to coordinate civilian war efforts, also purchased copies of the pamphlet to distribute through its member organizations—the Salvation Army, the Young Men's Christian Association (YMCA), the Young Women's Christian Association (YWCA), the National Catholic Community Services, the National Travelers Aid Association, and the National Jewish Welfare Board.

But a political storm soon erupted. In March 1944 the U.S. Congress launched an investigation into the pamphlet. A special committee of the House Military Affairs Committee, chaired by Representative Andrew J. May (Democrat, Kentucky), with Carl T. Durham (Democrat, North Carolina), Clifford Davis (Democrat, Tennessee), and Robert L. F. Sikes (Democrat, Florida), called Lieutenant Colonel Farlow, Professor Benedict, and Dr. Weltfish to testify. Representative May was particularly incensed by the assertion that northern blacks scored higher on intelligence tests than southern whites. The committee decided that *The Races of Mankind* was a subversive piece of Communist propaganda containing "many statements which range all the way from half-truths to innuendos to downright inaccuracies," and should not form part of the Army program.[77]

Two month later, in May 1944, Senator Theodore Gilmore Bilbo (Democrat, Mississippi), a segregationist who pursued a staunchly racist agenda throughout his tenure in the U.S. Congress, launched another attack against the pamphlet. Senator Bilbo feared the possible liberalization of immigration laws and saw *Races of Mankind* as part of a larger offensive to undermine the role of the white man in American politics and society. In a Congressional hearing, Bilbo declared:

There has never been a more disgusting conglomeration of scientific and physical facts than is to be found in this book. The whole scheme looks to the consummation of the plank in the platform of the Communist Party calling for the repeal of all laws forbidding the intermarriage of the races.... This same Communist Party advocated the immediate repeal of all immigration laws.... In other words, they propose to throw wide open the gates of this country to all classes, races, and nationalities of people.[78]

Bilbo believed that Boas, Benedict, and Weltfish were part of a Communist, Jewish, intellectual, cosmopolitan, and East Coast conspiracy that was threatening racial purity in the United States. The Congressional attacks succeeded in blocking the U.S. Army's use of *The Races of Mankind*. On March 10, 1944, the Army announced that the pamphlets that had already been bought would not be distributed. On August 25, 1944, the pamphlets were ordered destroyed. The USO similarly decided that its affiliates would no longer distribute the pamphlet.[79]

Despite this political storm, Reuther decided in 1946 to use *The Races of Mankind* as the basis for his animated short. The UAW-CIO was having trouble organizing in the South, where it was difficult to persuade blacks and whites to join the same union locals.[80] The film, Reuther hoped, would be an accessible treatment of race that could alleviate tensions.

The Brotherhood of Man was a sixteen-millimeter, eleven-minute short directed by Bob Cannon, animated by Ken Harris and Ben Washam, and written by Ring Lardner Jr., Maurice Rapf, John Hubley, and Phil Eastman. It was produced by Hubley, the inventor of one of the most successful characters in the history of animation, Mr. Magoo.[81] The film was previewed at the Museum of Modern Art in New York on January 3, 1946, and was well received. It was extensively reviewed by the press, and was shown in libraries and schools, and even considered as a potential American contribution at the Cannes Film Festival.

In occupied Germany, however, the film became the center of a secret foreign policy controversy. On May 20, 1947, Colonel W. E. Crist, the chief of the Area Intelligence Division, informed the War Department that the English edition of the *Ukranian Daily News,* the newspaper of the Ukranian Communist Party, had commented positively on *The Brotherhood of Man*. This insignificant item in a marginal newspaper alarmed American military intelligence. On July 1, 1947, Major General Daniel Noce, a fervent segregationist who had been promoted to head CAD that year, ordered that the film not be sent overseas. Records concerning the film were kept in a special file, classified as secret by CAD, that was not to be removed without the permission of the executive officer of the Special Staff Section of the War Department.[82]

The War Department's decision not to use the film in Europe infuriated Reuther, who wrote President Truman asking why the film was still not being shown in Germany. The film had been reviewed and approved by the theater commanders in Germany and Austria, endorsed by officials of the State Department and by many religious organizations, and shown to thousands of church groups and school groups in the United States, he argued. The only obstacle now was Major General Noce's bigotry:

[The film was] produced by the UAW-CIO as a public service and it propagandizes only for brotherhood, tolerance, and sympathy among human beings [but] at the present time, by order of Major General Daniel Noce, Chief of the Civil Affairs Division of the War Department, these films were sequestered in a warehouse and no use is to be made of them. I understand that this decision was made only because of the deep-seated personal prejudices of General Noce.[83]

This was the beginning of an interchange of letters between Reuther, the White House, and the War Department. Truman' secretary, William D. Hassett, wrote secretary of war Kenneth Claiborne Royall, asking his opinion as to whether the film should be distributed in Germany and Austria. Royall responded that showing the film "would be impolitic because of the controversy in the Congress, in the press, and elsewhere." He reminded the president of the 1944 Congressional investigations into *The Races of Mankind,* and claimed that General Noce "properly made the decision" not to use *The Brotherhood of Man*. Indeed, Royall regretted the fact that the previous CAD administration had purchased copies the film.[84]

Royall's refusal to show the short in Germany clashed with President Truman's policy on race. Truman was aware that the issue of race was eroding America credibility on human rights.[85] The President's Committee on Civil Rights, in a landmark report issued on October 29, 1947, argued that American racism was hurting America's reputation in the foreign policy arena. American journalists and religious leaders echoed this claim. Raymond Daniell, the former chief of the London bureau of the *New York Times,* wrote that that "the disfranchisement of Negroes in Southern states is cited as evidence of American hypocrisy when it comes to human freedom." He warned that "it would be a tragic blunder" if the United States did not fashion a new image in postwar Europe, distancing itself from the legacy of racism and showing the Europeans "what kind of people we really are." A new impression of America, Daniell reasoned, would help counter the Soviet's depiction of the United States. Since the Cold War was "a war of propaganda, of ideas, of ideology," the United States had to use the radio, press, and film as political weapons.[86]

In 1948 Reuther sent a second letter to Truman, claiming that *The Brotherhood of Man* was necessary to answer Communist propaganda in postwar Europe. Reuther argued that an antiracist film would help fight the anti-American onslaught. "Communist propaganda is steadily misrepresenting the motives of the American Government and its people," he wrote, and the "stereotype of America in which lynchings, discrimination, and Jim Crow predominates is being held before European workers." Reuther argued that the cartoon furthered the agenda of Truman's Committee on Civil Rights and presented an image of the United States that "would be welcomed by everyone engaged in the fight for the survival of democracy."[87] Reuther even offered to have the UAW-CIO distribute the film through trade union channels in Germany and Austria at no cost to the War Department.

The military, however, maintained its stance. Royall did not budge, and he supported the CAD position paper written by Colonel B. B. McMahon, deputy chief of the War Department's Reorientation Branch. In his memo, McMahon stated that he himself had reviewed the film, with William H. Draper Jr., undersecretary of the Army, and General George E. Eberle, and concluded that it was unacceptable because of its political message:

The sociological consideration on which the film was made . . . is essentially false—according to it, all races are created equal not only in strength, in brain matter and blood types, but subjected to like environment will respond similarly and in like degree. Environment is everything, and heredity meaningless. No cognizance is taken of such findings as those available to the Department of the Army, that wartime intelligence tests proved the definitive intellectual inferiority of American negroes as compared to American Whites. From a scientific viewpoint the film is misleading and its showing in the occupied areas would invite justified criticism from scientifically trained indigenous inhabitants as well as from members of Congress in the United States.[88]

Draper ordered that the film "be filed and not distributed."[89] In November 1948 Truman informed Reuther that, "guided by the policy laid out by the Department of the Army," the U.S. government had decided not to use *The Brotherhood of Man* in Germany and Austria.[90]

The controversy surrounding *The Brotherhood of Man* went beyond a dispute over screening an eleven-minute cartoon in postwar Germany and Austria. The episode illustrates the resistance of the War Department and the U.S. Army to the attempts to dismantle racist structures in the United States in the immediate postwar period. The Truman administration could not persuade Congress to pass antilynching laws and was struggling to desegregate the armed forces. Secretary Royall not only vetoed the film, but he systematically blocked President Truman's efforts to integrate the U.S. Army.[91]

The Brotherhood of Man was also caught up in the anticommunist mood of postwar America. In 1948 the House Un-American Activities Committee (HUAC) subpoenaed three of the film's writers, Hubley, Lardner (one of the Hollywood Ten), and Eastman. The California Senate claimed, in a report on "un-American" activities in its state, that *The Brotherhood of Man* was based on a pamphlet banned by the War Department, whose authors, Benedict and Weltfish, were affiliated with communist front organizations. According to the California report, the film was not attacked because of its content but rather because of the filmmakers's associations.[92]

The incident of *The Brotherhood of Man* is a small event in the history of the American military occupation of Germany, yet it illustrates how ideological preconceptions shaped American foreign policy. The profound racism prevalent among members of the U.S. Congress, the U.S. armed forces, and the War Department, continually hampered a rational response to Soviet anti-American propaganda in the early years of the Cold War. Moreover, the controversy suggests that the executive branch needed an agency that could counter Soviet anti-American propaganda without the constraints imposed by Congress and the armed forces. In order to contain and defeat the USSR, the United States needed to win the ideological confrontation of the Cold War. This struggle required a new level of intellectual and political flexibility, and the design and implementation of overt and covert programs of ideological penetration and propaganda that would have met considerable resistance in the United States.

Three

ICD's Blind Spot: The Fine Arts

Immediately after the cessation of hostilities, the Allies closed all German museums that had not been destroyed by air bombardment and ground fighting. Allied art experts canvassed the collections looking for art stolen by the Wehrmacht in occupied Europe and removed Nazi and militaristic art. Elsewhere, however, the German art scene revived quickly. German artists, alone or in small groups, showed their work in makeshift exhibitions in destroyed buildings. Viewing art became an important pastime for a population with few other recreational outlets. As soon as the Allies allowed the reopening of art galleries and museums, exhibitions were crowded, and lectures on art were immensely popular.[1]

The Soviets saw German artists as potential agents of social transformation and courted them as political assets. For the Communists, culture in general, and the fine arts in particular, was an important propaganda tool. Accordingly, SMAD established new cultural organizations, invited German intellectuals and artists to join, and offered them material incentives and rewards that were hard to reject in times of economic penury.

These Soviet efforts were not countered by an equivalent OMGUS program. ICD failed to appreciate the propaganda potential of the fine arts and did not participate actively in the German fine arts scene. This was not a trivial omission. The Nazis, aware of the German interest in culture and art, had exerted rigorous control over the fine arts and had used art as propaganda. The Nazi campaign against modern art had reinforced German cultural nationalism and isolationism. The aesthetic of the regime had left profound psychological marks on the German population, particularly young people educated during the Third Reich.

By 1946 the American military government's indifference to the fine arts in Germany had become a political liability. A small group of art experts in the Monuments, Fine Arts & Archives Section of OMGUS (MFA&A) recognized the mistake and, ignored by ICD, set out to design an American response to the SMAD campaign in the fine arts. These OMGUS cultural officers were of course constrained by American domestic politics. Modern art was anathema to socially conservative elements in the U.S. government, and therefore could not be made the official icon of American cultural freedom. Nonetheless, the MFA&A officers managed to incorporate the fine arts into the American reorientation program and facilitated the reemergence of modern art in Germany.

Germany and the fine arts

Since the 1840s, even prior to the creation of a unified Germany, German politicians had claimed that the fine arts symbolized the nation. In 1846 King Ludwig, speaking at the opening of the Neue Pinakothek, called for a nonelitist German art that would embody the national culture. From then on, the imperial German governments developed the artistic patrimony of the emerging nation by building art museums, promoting official art through a system of patronage, and exerting political control over fine arts collections and exhibits.[2]

Wilhelm II, who ruled from 1888 to 1918, considered himself the absolute authority on German aesthetics, much as Hitler would decades later. He was a religious mystic who loathed democracy, Jews, and socialism and associated modern art with all that he despised. As Kaiser, Wilhelm personally decided what could be shown in official exhibits. For him and other German social conservatives, modern art was a political danger, a challenge to the very foundations of the nation. Impressionism, in his view, was designed to subvert the moral education of the German people and undermine its Christian values. In fact, Wilhelm II denounced modern art as a sin against the German people.[3]

The campaign against modern art was inscribed within a conservative nationalist rhetoric that called for a "pure" identity grounded in a specifically German cultural tradition. The art historians Heinrich Wölfflin (1864–1945) and Wilhelm W. Worringer (1881–1965) argued for the existence of a German national style that superseded personal style and reflected "national character." Worringer maintained that artists expressed the "artistic will" of their ethnic group and that the artistic product was the result of racially determined factors. In 1893 the German cultural critic Max Nordau published *Degeneration,* which claimed that it was possible and necessary to separate "healthy" artists from "diseased" ones. Modern art was taken as the antithesis of German vigor, a degenerate product of the deranged and sick.[4]

Despite this campaign, German modern art flourished. German socialists denounced the imperial control of art as a form of political censorship. Modern artists excluded by the imperial *Kunstpolitik* organized the Berlin Secession Movement in 1892, and many German collectors and curators were instrumental in encouraging the development of modern art both in Germany and in France. In 1897 the Berlin National Gallery was the first museum to buy a Cézanne. Other German museums followed its lead. An outstanding example was the Folkwang Museum in Essen, which specialized in works by Gauguin and Van Gogh. Moreover, groups of modern artists such as Die Brücke and Blaue Reiter emerged in Germany, and journals like *der Sturm,* created in 1910, leapt to the defense of modern art.

After World War I, Berlin, Munich, Düsseldorf, and Cologne became the de facto capitals of French modern art. In 1919 the National Gallery in Berlin established a permanent wing dedicated to modernism. The two great patrons of Cubism, Daniel-Henry Kahnweiler and Wilhelm Uhde, were both German. And the total number of modern art works owned by German museums and private galleries reached some eighteen thousand.[5]

The Weimar avant-garde gave an extraordinary visual account of the horrors of World War I. German artists provided a trenchant visual record of the social and political tensions then plaguing Germany. In the 1920s some politically active German modern artists rejected abstraction and turned to a figurative pseudonaturalism inspired by daily life, projecting their ideological commitment into their work. George Grosz and Otto Dix, for example, denounced savage capitalism, rapid industrialization, and military defeat with sardonic brutality. The Verists, as they were called, viewed themselves as a political avant-garde with an aesthetic agenda, and even used the language of war to describe their attempts to change the world.[6]

This trend was incompatible with the Nazi project. Modernism implied individuality, political involvement, social criticism, and creative spontaneity, while Nazi ideology sought a regimented world devoid of political and social conflict. Worse, many avant-garde artists identified themselves with the extreme left. The Nazis saw such artists as political adversaries with competing views about culture, politics, and society, and equated modernism with everything they hated—Jewishness, cosmopolitanism, internationalism, and communism.[7]

The National Socialist regime further politicized the fine arts and placed Hitler, the Nazi Party, and the Nazi state as supreme arbiters of taste. Hitler had a fixation with both art and architecture, elevating them to the realm of the politically crucial. Most of his top functionaries were actively involved in the Third Reich's fine arts policy.[8] Although the Nazi ideologue Alfred Rosenberg, minister of propaganda and public enlightenment Joseph Goebbels, and the minister of education Bernhard Rust were officially in charge of cultural affairs, Hermann Göring, Albert Speer, Robert Ley, Wilhelm Frick, Walter Darre, Baldur von Schirach, Heinrich Hoffmann, and Martin Bormann also participated in the discussions about art and its role within National Socialism.

At first the Nazi potentates were deeply divided over the issue of modern art. All believed in the existence of a genetically determined Nordic art. They disagreed, however, on whether German Expressionism was an apt reflection of the German soul and the best symbol of the new order. Wilhelm Worringer, the art historian, maintained that German Expressionism was an ethnic product of the Nordic man. Goebbels agreed and wanted to make German Expressionism the cultural banner of the Nazi revolution. Alfred Rosenberg and the architect Paul Schultze-Naumburg, on the other hand, rejected all types of modernism as alien to the German cultural tradition. Rosenberg wanted to base the new Nazi order on old Teutonic ideals, and his Kampfbund für Deutsche Kultur (Fighting League for German Culture) was staunchly against modern art. Hitler ended the debate in 1934. After he dismantled the Sturmabteilung (Storm Troopers) and assassinated its leaders, including Ernst Röhm, Hitler announced the end of the National Socialist revolution. From then on, Hitler strove to consolidate the Nazi state, develop the German armed forces, and create a new German culture. And he decided that he would not tolerate any modern or antitraditionalist tendencies in art because they challenged his racial and national (*volkisch*) conceptions of culture.[9]

The Nazi campaign against modern art continued the Wilhelmian tradition of state control over the fine arts, but the Nazi regime's repressive policies were executed with a thoroughness and zeal possible only in a modern police state. Goebbels abolished independent art groups and created a single cultural organization controlled by the state, the Reichskulturkammer (National Chamber of Culture). The Reichskulturkammer left Rosenberg's Kampfbund without any political power. Modern artists could not work unless they changed their style and were accepted by the new organization. By 1935, the Reichskulturkammer had 100,000 members, including 14,300 painters and 2,900 sculptors. The Nazis purged the art world of Jews and Communists, eliminated independent art criticism, and eventually removed modern art from museums and galleries.[10]

The campaign against modern art and modern artists gained intensity after the 1936 Berlin Olympic Games. In 1936 Adolf Ziegler, the painter who served as the head of the Reich Chamber of Visual Arts, directed a massive exercise in iconoclasm and purged German museums and galleries of modern art. Ziegler and his assistants visited a hundred German museums and requisitioned thousands of paintings, drawings, and sculptures. Six hundred fifty of the requisitioned works were selected for the *Entartete Kunst Ausstellung* ("Degenerate Art Exhibit"), which opened on July 19, 1937, in the Archeological Institute in Munich. The exhibit included work by internationally renowned German artists, with paintings and sculpture representing all modern styles—Expressionism, Verism, Cubism, Impressionism, Abstraction, Bauhaus, Dadaism, and New Objectivity. Alongside these modern works were photographs of people with severe deformities and art made by mental patients. This juxtaposition stressed the Nazi interpretation of modern art as degenerate, decrepit, decadent, and destructive to the purity of German culture. Modern art was labeled as Jewish, anti-German, antinational, international, cosmopolitan, and Communist. It was the embodiment of foreignness—rootless, gross, dirty, incomplete, brutal, confused, and nihilistic.[11]

By 1938, when the purge of modern art was declared complete, the regime had confiscated all modern art from German museums and galleries. This policy of censorship and suppression was more than a reflection of Hitler's distaste for modernism. It was a deliberate political attempt to isolate the German public from foreign cultural influences in order to create a hegemonic Nazi cultural narrative.

Simultaneously, the Nazis developed a positive fine art policy that integrated painting and sculpture into their propaganda apparatus. The Nazis claimed that German art had to reflect the German soil, blood, and spirit, to reveal the racial purity of the German *Volk* and mirror the greatness of "Aryan" culture. The Third Reich commissioned, sponsored, and exhibited "New German Art," the official art of the regime. In general terms, New German Art did not differ much from the official art of other dictatorships of the twentieth century—it was a figurative neoclassical style with a restricted portfolio of themes and genres aimed at propagating the cult of the authoritarian system and its leader. In Germany, this allegorical and emblematic art reinforced the Nazi ideological tenets of nationalism, ethnocentrism, racism, and conformity, and depicted a fantasy world devoid of suffering, conflict, and social contradictions.

Hitler played the role of the great patron of the arts of the Third Reich and made a point of attending the openings of all the important art exhibits, which were filmed and widely publicized. He systematically bought works of New German Art for the Reich Chancellery and for his private residences. In 1937 Hitler organized the first *Große Deutsche Kunstausstellung* ("Great German Art Exhibit"). This exhibit, a showcase of New German Art, opened in the newly constructed Haus der Kunst (House of German Art) in Munich, and took place at the same time as the "Degenerate Art Exhibit." Hitler institutionalized the *Kulturtagung* (cultural conferences) as part of the yearly Nuremberg rallies and established a National Day of German Art. He commissioned the building of new museums and paraded their maquettes through German towns like religious icons. The government enforced extensive media coverage of New German Art exhibits and produced high-quality fine art magazines, such as *Die Kunst im Deutschen Reich* ("Art in the Third Reich"), *Kunst* ("Art"), and *Kunst dem Volk* ("Art of the People").[12] Goebbels organized art exhibitions in factories, and integrated New German art into mass culture through films, posters, postcards, and advertisements. New German Art and Nazi paraphernalia saturated the visual sphere. By 1937 the fine arts budget was the largest it had ever been in Germany.[13]

Hitler thought that the elimination of "degenerate" art would be automatically followed by the emergence of an extraordinary "Aryan" art. He believed that Nazi artists could continue the tradition of the nineteenth-century German painting that he revered. Yet both he and Goebbels were disappointed with the results and felt that the neoclassical style that emerged was mediocre at best. The *Führer* was not pleased with the artists of the Third Reich, and Goebbels called the New German Art "Munich-school kitsch."[14] Nonetheless, the Nazis used New German Art as political propaganda, and blanketed Germany with this artwork. The 1942 *Große Deutsche Kunstausstellung* attracted 846,674 visitors.[15] These numbers reflect a political reality that was totally lost on OMGUS and Washington: in Nazi Germany, the fine arts mattered.

ICD and the German fine arts

JCS 1067 did not mention German cultural revival and did not stipulate a definitive cultural policy for Germany. The Americans were certainly aware of the political importance of the mass media in propaganda and mass persuasion, but they did not attach political relevance to the fine arts. ICD's involvement in the fine arts was purely negative and limited to denazification. ICD did not even have a section dedicated to the fine arts; the personnel of its Theater and Music Section were in charge of monitoring the fine arts and vetting painters and sculptors who had a Nazi past. Only artists who completed political questionnaires and were exonerated from the taint of Nazism were authorized to register as artists, to have access to art supplies, and to exhibit. The Theater and Music Section also licensed galleries and art journals, authorized art shows and art deals, and controlled exhibition brochures and posters.[16]

Yet ICD did not have a positive fine arts policy. The OMGUS fine arts experts were not in ICD but in the Monuments, Fine Arts & Archives Section (MFA&A), part of the Restitution Branch of the Economics Division. MFA&A was created in 1944 following an appeal to President Roosevelt by the American defense Harvard Group and a committee of the American Council of Learned Societies in New York, asking for the protection of art work in the European Theater. Roosevelt created the American Commission for the Protection and Salvage of Artistic and Historic Monuments, known as the Roberts Commission because it was headed by Supreme Court Justice Owen J. Roberts. The Roberts Commission, a link between the War Department and the art historical community in American academia, provided the U.S. Army with the "Supreme Headquarters Official List of Protected Monuments," a survey of European cultural icons that should be preserved. On May 26, 1944, just before the invasion of northwest Europe, General Eisenhower signed a letter to his generals explaining that the area had monuments of great cultural importance, and that it was Allied policy to attempt to protect cultural treasures. From that point on, a monuments specialist officer, or "monument man," was attached to the staff of each army commander.

Most MFA&A officers were art historians, architects, museum curators, art professors, sculptors, or painters. Their job was a difficult and dangerous one: to move with the fighting troops and warn them that certain cultural relics and structures—such as cathedrals—should not be destroyed or looted during combat. The monuments men also checked and recorded war damage to historical buildings, museums, fine art collections, and libraries. As Marvin C. Ross, a monuments man, later recalled, "The protection of cultural material was woven completely into the armed forces operating under SHAEF."[17] Yet, at the end of the war, MFA&A was one of the smallest outfits of the Allied armies—in autumn 1945 it had eighty-five members.

After V-E Day, MFA&A was dedicated to the search and restitution of art stolen by the Nazis in France, Holland, Belgium, Czechoslovakia, Russia, Poland, and Germany. The process involved a good deal of detective work.[18] Finally, MFA&A officers oversaw monument repair and reconstruction, the financial regulation of art trade, the management of the Berlin art museums, and the administration of German libraries. However, MFA&A was barred from intervening in the postwar German art scene.[19]

MFA&A was not meant to be an instrument of cultural policy. This seems curious, given the fact that its personnel included such first-rate art historians and art scholars as Craig Hugh Smyth, former curator of the National Gallery of Art, James Rorimer, former curator at the Cloisters, Edith Appleton Standen, former curator of the Widener Collection at Harvard University's Fogg Museum, and Lincoln Kirstein, a collaborator of Alfred Barr, the director of the Museum of Modern Art in New York. Some MFA&A officers did establish strong personal relations with German artists and art scholars, yet they did so only in their private capacity, not as members of OMGUS. According to Standen, the work of the MFA&A officers was hampered by the "indifference and even disapproval of their cultural activities by higher authority in Military Government, in striking contrast to the official policy in the fields of music, the theater, radio, the cinema, etc."[20]

The fine arts in the Soviet zone and sector

The Soviet attitude toward the fine arts was drastically different. The Communists regarded all aspects of culture, from the mass media to the fine arts, as intrinsically political and susceptible to exploitation as direct or indirect propaganda. A decree on the restoration of cultural life in the Soviet zone established without ambiguity that SMAD would have political control of art exhibitions in Eastern Germany. It called for the elimination of Nazi racist and militarist doctrines, and for the "mobilization of art as part of the struggle against Fascism and of the re-education of the German people in the spirit of true democracy." SMAD considered the revival of the German fine arts a political task and organized a multilayered system of control to manage it.[21] From 1945 to 1949, SMAD, coordinating its activities with the German Communist Party (KPD) and later the Socialist Unity Party (SED), strove to attract the German cultural intelligentsia through material incentives, threats, and direct coercion.

In Berlin, the Soviets hosted their first cultural event even before the Inter-Allied Military Kommandatura assumed control over the city. The Soviets established the Kammer der Kunstschaffenden (Chamber of Art Workers) in Berlin in June 1945 and a few months later created the Kulturbund zur demokratischen Erneuerung Deutschlands (Cultural League for the Democratic Revival of Germany), which sponsored art exhibits and other cultural events throughout Germany. Simultaneously, the Soviet military government established the Deutsche Verwaltung für Volksbildung (German Agency for Popular Enlightenment) with a subdivision in charge of fine arts and museums.[22]

The Soviet fine arts policy in postwar Germany had two phases. From 1945 to mid-1946, SMAD organized exhibitions of German art that included works by modern artists labeled "degenerate" by the Nazis. This period of aesthetic tolerance and openness ended when, in August 1945, Stalin ordered Andrei Zhdanov, the third secretary of the Soviet Communist Party, to launch an offensive against cosmopolitanism and modernism in the Soviet Union. Although the Zhdanov decrees were not published in Soviet-occupied Germany, SMAD art policy changed markedly. From August on, Soviet cultural officers began to demonize modern art and to glorify Socialist Realism.

In August 1946 SMAD opened the *Allgemeine Deutsche Kunst Ausstellung* ("German Art Exhibit") in Dresden. The largest show of modern art in Germany since the *Entartete Kunst Ausstellung* of 1937, it included works by such artists as Ernst Barlach, Max Beckmann, Otto Dix, Lyonel Feininger, Conrad Felixmüller, Erich Heckel, Ernst Ludwig Kirchner, Paul Klee, Oskar Kokoschka, Käthe Kollwitz, Wilhelm Lehmbruck, Max Pechstein, and Karl Schmidt-Rottluff. Paradoxically, the *Allgemeine Deutsche Kunst Ausstellung* marked the beginning of the SMAD campaign against modern art. The highest-ranking political intelligence officers in SMAD, General Tiul'panov and Lieutenant Colonel Dymschitz, attacked what they called "formalism" in art and denounced modern artists. It is worth noting that Hitler also launched his antimodernist campaign with a modern art exhibit. A few days after the inauguration of the *Allgemeine Deutsche Kunst Ausstellung,* Dymschitz published the first of a series of seven articles in the *Tägliche Rundschau* in which he explained why Socialist Realism was an important tool for the democratic reconstruction of Germany.[23]

German response to the *Allgemeine Deutsche Kunst Ausstellung* in fact bolstered the Soviet campaign against modern art. Three-fourths of the visitors to the exhibit claimed to have disliked what they saw.[24] SMAD exploited this resistance to modern art—a direct consequence of twelve years of Nazi indoctrination—to advance Socialist Realism as an alternative style that reflected an authentically antifascist and democratic message. The first cultural conference organized by SED, which coincided with the Dresden exhibit, gave party officials the opportunity to demand a German art that was "democratic" in content and national in form. Echoing the Soviet line on modernism as a sign of Western decadence, SED denounced "formalism" and the "reactionary" tendencies of the German art in the American zone.[25]

Modern art and the American government

In 1946, while Stalin and Zhdanov launched the final attack against modernism and modern artists in the USSR and SMAD began its offensive in Germany, the anti–modern art movement reached its peak in the United States. An alliance that included influential members of the U.S. Congress, the Hearst newspapers, and conservative art associations denounced American modern art on aesthetic, political, and moral grounds. The antimodernist groups defined modern art as cosmopolitan, foreign, anti-American, and Communist, and deemed it ridiculous, subversive, and threatening to the American way of life.[26] This conservative alliance was strong enough to derail the first American project of cultural diplomacy.

In January 1946, six months after the end of the war, a group of diplomats in the State Department decided that a victorious United States had to take the lead in cultural matters on an international scale. The Fulbright Act inaugurated an international exchange program, and in spite of the opposition of the House of Representatives, assistant secretary of state William Benton persuaded the Senate to appropriate nineteen million dollars for the newly formed Office of International Information and Cultural Affairs (OIC). The function of the OIC, a peacetime propaganda agency, was to show that the United States had emerged from World War II not only as the military leader of the Western world but as the cultural leader.[27]

To make this case, Benton organized a "world-wide art program" to publicize American accomplishments in the fine arts. Benton saw modern art as a vibrant symbol of American cultural achievement, suitable to be displayed abroad. He appointed J. Leroy Davidson, former curator of the Walker Art Center in Minneapolis, to serve as visual arts specialist in the Division of Libraries and Institutes of the State Department.[28] Davidson conceived the idea of sending an exhibit of modern American paintings and drawings, *Advancing American Art,* to tour Europe and South America under government sponsorship. He acquired seventy-nine oils and seventy-three watercolors by forty-five well-known American artists, for a total of forty-nine thousand dollars. The collection—which included works by Stuart Davis, John Marin, Georgia O'Keeffe, Ben Shahn, Ben Zion, and George Grosz, who had left Germany and become a U.S. citizen—was first shown to the American public at the Metropolitan Museum of Art, in New York City, in 1946. The show did not include works of the New York school of Abstract Expressionism. Nonetheless, it was praised by many leading American art critics.[29]

But there were also vicious political attacks. In October 1946, as *Advancing American Art* began its transatlantic tour, meeting with critical success in Europe and South America, the Hearst papers and conservative art groups started a defamatory campaign, alleging that the show was aesthetically and politically dangerous. The art critic of the *New York Journal* wrote that the exhibit was "not American at all"; rather, its "roots" were "in the alien cultures, ideas, and sickness of Europe."[30] Assessing the flood of animosity, Marilyn Robb, writing in *Art News,* commented:

Although it had been originally feared . . . that the abstractions in the collection would cause criticism, the representational works aroused most of the fury. It was Kuniyoshi's mild *Circus Rider* which elicited from President Truman the remark "If that's art I'm a Hottentot."[31]

The Los Angeles Art Club claimed that the show was "inimical to our form of government." The American Artists Professional League, an organization of "modern classicists," claimed that the art selected for the exhibit was contaminated by influences "not indigenous to our soil."[32]

Members of the new, Republican-dominated Congress alleged that the show depicted a warped image of American society. George Dondero, the Republican senator from Michigan who led the campaign against the exhibit, claimed that "modern art was communism," and thus fundamentally un-American.[33] Dondero argued that American painting should represent a strong, healthy nation, whereas *Advancing American Art* suggested a contorted and deformed image. The show, he claimed,

is communistic because it is distorted and ugly, because it does not glorify our beautiful country, our cheerful and smiling people, and our great material progress. Art which does not portray our beautiful country in plain, simple terms that everyone can understand breeds dissatisfaction. It is therefore opposed to our government, and those who create and promote it are our enemies.[34]

Representative Fred Busbey (Republican, Illinois) echoed Dondero's views. The exhibit did "not represent American culture," he said. It was "as foreign to the American way as is the Moscow radio" and had harmed America's image abroad:

The so-called art exhibit that was sent abroad by the State Department is a disgrace to the United States. . . . The alleged art exhibition has sinister aspects. The ... trashy paintings sent on tour by the State Department ... have done our country harm abroad. Foreigners must be wondering what kind of crackpots assembled such a jumble of paintings.[35]

Representative John Taber (Republican, New York), chairman of the House Committee on Appropriations, chastised the State Department for conducting an "anti–Cultural Relations Program." He claimed the paintings in *Advancing American Art* were "a travesty upon art" and made "the United States appear ridiculous in the eyes of the foreign countries," promoting "ill-will towards the United States."[36]

Assistant secretary of state Benton was summoned by the House Appropriations Committee, and Congress blocked further funds for the Cultural and Information Program. President Harry Truman, under political pressure, expressed his disapproval of the exhibit, saying that the paintings looked more like scrambled eggs than art. The State Department was instructed to withdraw the exhibit. On April 4, 1947, a front-page article in the *Los Angeles Examiner,* reflecting the conservative equation of modernism and subversion, exulted, "Radical Art Tour Halted."[37]

The cancellation of the tour likewise elicited protests. Ninety leading artists, art dealers, museum directors, art critics, and art scholars sent a public letter to President Truman and secretary of state George C. Marshall denouncing the decision. Another letter bearing a thousand signatures was sent to Truman, members of Congress, and the secretary of state. Nonetheless, Secretary of State Marshall decided that "no more taxpayers' money" would go toward promoting modern art.[38]

Twenty months after the inaugural Metropolitan Museum exhibit, the paintings and drawings in *Advancing American Art* were shown in the Whitney Museum in New York, this time as government surplus property, offered for sale at below-market prices by the War Assets Administration.[39] The artists who had participated in the show were all investigated by the House Un-American Activities Committee. The State Department fired Davidson, who joined the art history faculty at Yale University, and his post was abolished. Not until 1953, when the United States Information Agency was established, did the U.S. government again sponsor American art exhibits abroad.[40]

The fate of *Advancing American Art* illustrates the intensity of the battle to define which American cultural icons would represent the United States abroad. The fragile national political alliance President Roosevelt had constructed to win the war had collapsed, and political power had shifted to the isolationist, racist, and ultraconservative right of the Democratic and Republican Parties. The new masters of the U.S. Congress had no qualms about exporting their notions of democracy, capitalism, and morality through the mass media, and especially through Hollywood. They did not object when American popular culture took Germany by assault, generating revenue for American companies. Yet they blocked the effort to export modern art as an exemplar of American postwar power.

The direct intervention of the U.S. Congress in cultural matters complicated and limited American response to the Soviet antimodernist campaign in Germany. With Congress censoring modern art and harassing modernist artists, it was difficult to denounce the USSR for doing exactly the same. It is not unreasonable to assume that many in the Pentagon, the State Department, and the upper echelons of OMGUS agreed with the congressional appraisal of modernism. In any case, OMGUS relied on Congress for appropriations, so a defense of modern art in Germany would have been politically unwise.

A new role for MFA&A

The battle over *Advancing American Art* illustrates to what extent Congress was a serious hindrance in fighting the cultural Cold War. An isolationist, xenophobic, and ultraconservative majority effectively blocked U.S. efforts to counter Soviet anti-American propaganda. In Germany, cultural operations to promote modern art therefore had to be covert in order to avoid congressional scrutiny.

In 1947 SMAD radicalized its campaign against modern art, abandoning all pretense of ideological neutrality. In November Lieutenant Colonel Dymschitz delivered a public lecture at the Soviet Haus der Kultur in Berlin, on the theme "Soviet Art and Its Relation to Bourgeois Art." Dymschitz called for an art for the people and of the people, understandable by all. German antifascist artists and intellectuals, following the Soviet model, were encouraged to depict "reality"—the life and struggle of the working class—in the context of class warfare, and to contribute to a new, revolutionary, and anticapitalist society founded on progressive social ideals. Surrealism and abstraction were deemed worthless, individualistic, capitalist, bourgeois, and decadent. A year later, Dymschitz wrote a series of articles in the *Tägliche Rundschau* critiquing formalist, decadent, and reactionary tendencies in German art.[41]

Still, ICD did not address the Soviet antimodernist offensive. Nor did it design a full-fledged campaign to support the reemergence of German modern art. However, two MFA&A officers, Hellmut Lehmann-Haupt and

Edith Appleton Standen, realized that antimodernism was politically dangerous in postwar Germany. On the one hand, it was a powerful cultural remnant of Nazism; on the other, it was at the heart of the Soviet rhetoric. Lehmann-Haupt and Standen were able to persuade the MFA&A leadership of the importance of developing an art policy for the American zone and sector.

Lehmann-Haupt (1903–1992) was born in Berlin, the son of Carl F. Lehmann-Haupt, a professor of ancient history at Innsbruck, and Therese Haupt, a playwright. He attended universities in Berlin and Vienna, majoring in fine arts and minoring in German and English literature. In 1927 he received his PhD from the University of Frankfurt and became an assistant curator at the Gutenberg Museum in Mainz. He left that position in 1929 when he emigrated to the United States, where he worked as indexing editor for the *Encyclopaedia Britannica* and proofread for the New York editorial house Marchbanks. In 1930 Lehmann-Haupt became curator of the Rare Book Department of the Columbia University Library, and in 1938 he was appointed assistant professor of book arts in the School of Library Services. During World War II, Lehmann-Haupt served in London (1944–1945), first as deputy chief of the German Policy Desk of the Office of War Information, and later as psychological warfare officer at SHAEF. After the defeat of Germany, Lehmann-Haupt was transferred to Berlin as a civil art liaison officer for MFA&A (1946–1948) and later became an art intelligence coordination officer (1948–1949).

In October 1946 Lehmann-Haupt urged the quick implementation of a coherent OMGUS fine arts policy for Germany to be carried out by the MFA&A. He proposed a policy aimed both at encouraging non-Nazi German artists and art scholars, and at consolidating strong cultural and affective ties between them and the American art establishment. In his October 1946 memo, "MFA&A Functions," Lehmann-Haupt outlined a plan to transform MFA&A into the catalyst of an artistic revival in the American zone and sector. In addition to its main existing responsibilities—searching for looted art and managing the collecting points for recovered work—Lehmann-Haupt added a series of new political functions: sponsoring exhibitions of contemporary German and American art, encouraging German artists, and organizing the retraining of German museum personnel and art historians.[42]

Lehmann-Haupt understood the German postwar context in ways that escaped his American colleagues in ICD. He was interested in the connection between authoritarian politics and art. He followed the development of antimodernist cultural agendas in European and Latin American populist

Generals Dwight D. Eisenhower, Omar N. Bradley, and George S. Patton, and other high-ranking U.S. Army officers, inspect the charred remains of prisoners at the newly liberated Ohrdruf concentration camp, April 12, 1945. United States Holocaust Memorial Museum, Washington, D.C.

American congressmen view open ovens in the Buchenwald crematorium, April 24, 1945. Pictured, from left to right, are Senator Alben W. Barkley (Democrat, Kentucky), Representative Edouard V. Izac (Democrat, California), Representative John M. Vorys (Republican, Ohio), Representative Dewey Short (Republican, Missouri), Senator C. Wayland Brooks (Republican, Illinois), and Senator Kenneth S. Wherry (Republican, Nebraska). United States Holocaust Memorial Museum, Washington, D.C.

American editors and publishers examine the corpses of victims at Dachau concentration camp, May 4, 1945. United States Holocaust Memorial Museum, Washington, D.C.

German civilians, under U.S. military escort, are forced to view corpses in the newly liberated Buchenwald concentration camp, April 16, 1945. United States Holocaust Memorial Museum, Washington, D.C.

An American GI monitors German civilians as they view corpses at Flossenbürg concentration camp, Namering, May 17, 1945. United States Holocaust Memorial Museum, Washington, D.C.

Billboard advertising the OMGUS-produced atrocity film *Todesmühlen* at the Luitpoldtheater in Munich, 1946. Haus der Bayerischen Geschichte, Augsburg.

Visitors view photomurals showing the liberation of Bergen-Belsen concentration camp at the "Lest We Forget" exhibition, Library of Congress, Washington, D.C., June-July 1945. United States Holocaust Memorial Museum, Washington, D.C.

Adolf Hitler, Karl Kolb, Eugen von Schobert, Adolf Ziegler, Joseph Goebbels, Wilhem Frick, and Bernhard Rust, at the Haus der Deutschen Kunst, Munich, July 10, 1938. Bayerische Staatsbibliothek, Munich.

Iconoclasm in occupied Germany. The Nazi eagle that once adorned the facade of the Haus der Deutschen Kunst, in Munich, lies decapitated, 1945. Gallery Archives, National Gallery of Art, Washington, D.C.

(Left) Lieutenant Colonel Alexander L. Dymschitz, head of the cultural division of the Office of Information, Soviet Military Government in Germany, Berlin, 1947. Bundesarchiv Koblenz.
(Right) General Sergei I. Tiul'panov, director of the Office of Information, Soviet Military Government in Germany, Berlin, 1946. Landesarchiv Berlin.

(Left) General Lucius D. Clay, military governor, Office of Military Government U.S. in Germany, 1947. Harry S. Truman Library, Independence, MO.
(Right) Brigadier General Robert A. McClure, the first head of the Information Control Division, Office of Military Government U.S. in Germany, photographed in August 1948, as chief of the New York Field Office of the U.S. Army's Civil Affairs Division. Robert D. McClure Photo Collection.

Henry Koerner, *The Prophet*, 1945–1946, oil painting on Masonite, shown in the exhibit *Ausstellung Henry Koerner U.S.A. 1945–1947* at the Haus am Waldsee, Berlin. Mrs. Bernard Grossman Collection.

(Left) Dr. Hellmut Lehmann-Haupt, former Monuments, Fine Arts, and Archives officer, Office of Military Government U.S. in Germany, in the 1960s. Christopher Lehmann-Haupt Photo Collection.

(Right) Captain Edith A. Standen, Monuments, Fine Arts, and Archives officer, Office of Military Government U.S. in Germany, at Wiesbaden Collecting Point, April 19, 1947. Gallery Archives, National Gallery of Art, Washington, D.C.

President Harry S. Truman, Generals Dwight D. Eisenhower and Omar N. Bradley, and secretary of war Kenneth C. Royall, February 7, 1948. Harry S. Truman Library, Independence, MO.

Film stills from *The Brotherhood of Man*, Hollywood Quarterly, July 1946.

Herbert Sandberg, cofounder of *Ulenspiegel*, in 1970. The Granger Collection, New York, NY.

Günther Weisenborn, cofounder of *Ulenspiegel*, in 1959. Deutsches Literaturarchiv, Marbach.

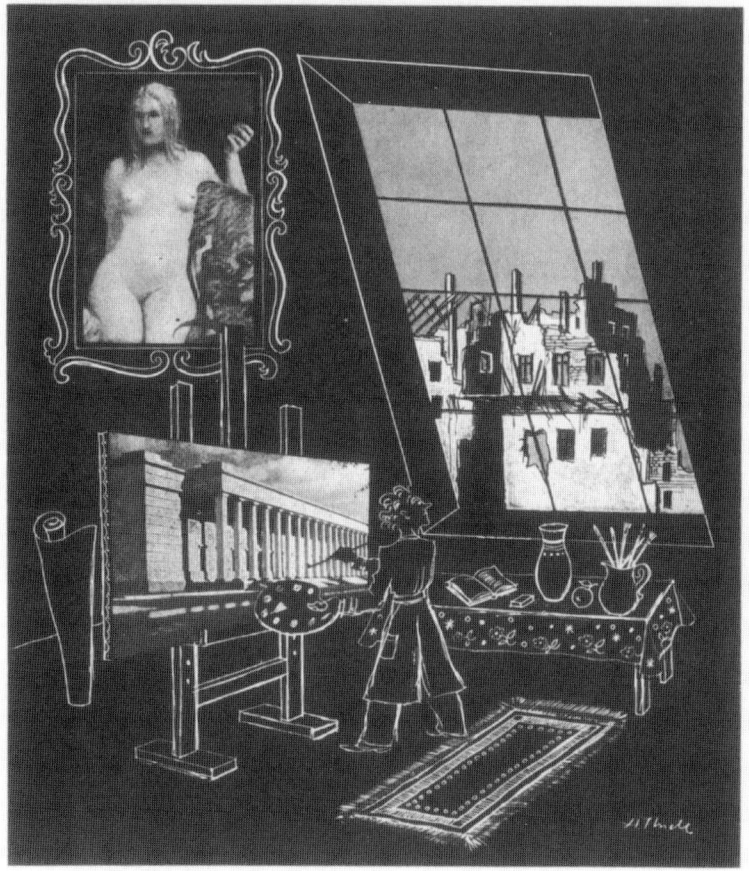

Herbert Thiele, "Sleeping Beauty, wake up! Hold on, wasn't the Thousand Year Reich just there?" *Ulenspiegel* cover, year 1, no. 1, December 1945.

Heinrich Kilger, "If we had won the war." *Ulenspiegel* cover, year 1, no. 4, February 1946.

Karl Holtz, "He got it made, he lives under a better sky." *Ulenspiegel* cover, year 1, no. 24, November 1946.

Elizabeth Shaw, "Land-Land!, Deutsch-land!" *Ulenspiegel* cover, year 4, no. 23, December 1949.

dictatorships and was keenly aware of the Nazi campaign against modern art. Moreover, Lehmann-Haupt appreciated the importance of the fine arts in German life and realized that the Nazi leaders had capitalized on this tradition, using the fine arts as a means of reeducation and mass propaganda.[43] In a November 1947 memorandum to Richard Howard, the head of MFA&A, he explained the political importance of the Nazi art policy:

> Control was only partially a negative tendency, in so far as it concerned the elimination of undesirable individuals from the participation in national life, the eradication of trends and ideas considered contrary to Nazi ideology (Entartete Kunst) and the appropriation of Jewish art property. Positively, the Party and the Government saw a powerful weapon in the political activation of practically all phases of art life. The Propaganda ministry and several separate agencies recognized with a shrewd instinct and considerable insight into the structure of the German mind the enormous importance of cultural matters in the life of the nation.[44]

Lehmann-Haupt clearly understood that the Third Reich had used the fine arts as propaganda and indoctrination:

> The annual exhibitions of the "Haus der deutschen Kunst" were powerful instruments in the glorification of the Nazi leaders, of militarism and the propagation of various Nazi doctrines. An elaborate system of governmental control instilled Nazi ideologies into practically all phases of art life.... In other words, museums, exhibitions and the very process of artistic creation were made into a powerful medium of education and public information.[45]

By suppressing modern art and modern artists, the Nazis had severed the cultural ties between Germany and the rest of Europe, eliminating artistic influences that they considered "foreign" and reinforcing the German public's growing cultural isolation.

Lehmann-Haupt's arrival in Berlin in 1946 coincided with the intensification of SMAD's campaign against modern art. Lehmann-Haupt attended the Dymschitz lectures on Socialist Realism and witnessed the Soviet cultural officer's attack on surrealism and abstraction. According to Lehmann-Haupt, Dymschitz claimed that "contemporary bourgeois art" was "decadent, individualistic, capitalistic," clashing with "genuinely socialistic and democratic art" that carried "the message of social progress."

SMAD's new antimodernist agenda and the Soviet preoccupation with the political control of German art led Lehmann-Haupt to conclude that "the Soviet dictatorship [was] as vitally concerned with the arts as was the Nazi state."[46]

Years later Lehmann-Haupt recalled his concern at the American indifference toward the German art scene:

Those of us who went to Germany with the United States Military Government after the war could not help but feel a little envious, at first, of the cultural program of the Soviet authorities in their zone of occupation. Some of us were dismayed to find that the Washington policy makers apparently had never heard of Nazi art policies and had literally no idea of the deep impression the Nazis had made. Our Monuments, Fine Arts, and Archives Section was strictly limited to matters of cultural property.... We were warned off the premises of cultural reorientation in no uncertain terms. There was at that time no provision for any branch of the United States Military Government to apply the general directives for democratic reorientation in the field of the arts.[47]

In February 1947 Lehmann-Haupt explained the political importance of art in Germany and why he thought the fine arts should be part of the American "cultural reorientation" program:

German museum and art life is not, as it may seem at first sight, a neutral field, devoid of political implications. On the contrary it is an active and vital factor capable of exerting considerable influence on public opinion and in formulating emotional and psychological attitudes. It must therefore be considered an important educational factor and, at least to some degree, a medium of public information. The authority and prestige which all manifestations of cultural life enjoy in the German community is very considerable. It is therefore necessary to realize that the activities of museums and other cultural and artistic organizations in Germany have a much greater influence, compared with other manifestations of public life, than in the United States or Great Britain.[48]

The revival of the German art scene, then, was essential to the American reorientation program.

In his capacity as cultural officer, Lehmann-Haupt had the opportunity to interview many of the leading figures of German art living in the American zone and sector, and he was alarmed by the persistence of extreme cultural nationalism and antimodernism. The defeat of the Third Reich had not brought an end to German cultural insularity. Art history professors continued to discourage students from studying modern art; modern artists struggled in vain to get commissions from German authorities, and both private and public innuendos against modern art were commonplace.[49]

The rejection of modern art, Lehmann-Haupt felt, was the most reliable indicator of reactionary nationalism. A positive American fine arts policy was needed to counter the "unhealthy return to cultural isolationism and inbreeding" that was compromising the American democratization mission.[50] In an undated essay, "German Art Today," Lehmann-Haupt reflected on the connection between antimodernism and Nazism:

> The all-important question seems to me, in what direction indifference and hostility towards modern art in Germany is moving today, whether it is going along at a steady pace or gaining volume and sharpness. Many observers on the scene with whom I have talked, insist that these unfavorable reactions, ranging from indifference to ironical condescension and outright hostility, are no different in Germany from Philistine attitudes anywhere else in the world. Such complacency does not seem to me justified. The deliberate and prolonged activation by the Nazis of all such attitudes into a political dogma of the first magnitude has undoubtedly had a profound effect. I find it impossible to believe that such attitudes have survived independently from association with the Nazi ideology. Today they are still perhaps in rear-guard. However, in connection with the steady rise of neo-fascist political organizations in Western Germany they could very quickly assume a different meaning and importance. They could lead right back into a situation of totalitarian tyranny.[51]

The rejection of modern art, Lehmann-Haupt concluded, indicated a failure of the American denazification and reorientation effort.

Lehmann-Haupt continued to propose "methods for a reorientation program" aimed at consolidating cultural relations between the United States and Germany. In March 1947, after a visit to Bavaria, he suggested that the MFA&A should do more to combat German cultural nationalism and to widen the German worldview. He recommended (a) creating exchange programs for American and German artists, art scholars, and art students; (b) publishing the work of contemporary German artists; (c) organizing lectures and discussions on art; (d) sending American art publications to Germany; (e) bringing American art experts to lecture in Germany; and

(f) sponsoring periodic art conventions for German art administrators and art professors.[52] By June, however, Howard still did not have OMGUS authorization to involve MFA&A personnel in overt art policy. The most he could do was to allow MFA&A officers to participate in German art life in an informal capacity. In a letter to Frederick A. Sweet, a curator at the Art Institute of Chicago, Howard wrote:

> Privately every member of the Section, both at OMGUS and in the Zone, is doing a good deal to encourage the artists and help them.... Among other things, several of us have purchased artists' supplies from the States out of our own pockets and distribute to artists who need them. We religiously attend exhibitions, even though many of them are pathetic shadows of contemporary art activity. We semi-officially advise the officials of the various German governments, the artists' associations, and the museum men, even when we are not officially permitted to do so.[53]

General Clay also wrote to Sweet in June 1947, outlining some planned actions in the fine arts: ICD and the Reorientation Branch Work Program of the War Department's Civil Affair Division were planning exhibitions of American Art, filmstrips, and lectures "with the aims of stimulating German artistic endeavor and increased contacts between German and American art circles." However, he explicitly downplayed the importance of American art policy in Germany: "It must be realized that housing and food necessarily come before exhibition space and artistic materials."[54] Clay did not perceive the fine arts as instruments of politics and propaganda, and OMGUS continued to ignore the advice of its art experts.

Captain Edith Appleton Standen, the director, from August 1946 to August 1947, of the MFA&A art collecting point at Wiesbaden, was troubled by the situation. Standen (1905–1998), born in Halifax, Canada, had become a U.S. citizen in 1942 and joined the Women's Army Corps. In 1945 she was assigned to MFA&A and stationed first in Frankfurt, where she inventoried the Reichsbank collection. On November 12, 1947, after leaving her post in Wiesbaden, Standen wrote a personal letter to Craig Hugh Smyth, former director of the Munich Collecting Point, describing the obstacles facing the development of a coherent fine arts policy in Germany: "Maybe I've been too critical of ICD, but the situation is ridiculous.... The ICD people in Germany have no interest in our fields."[55] Standen authorized Smyth to use her memo in a meeting with General William Draper, who served as Clay's economic advisor:

If General Draper should ask you what practical step you want him to take, I should suggest a cable, from him or his Economics Division chief, to Econ. in OMGUS, along these lines: "Cultural activities of MFA&A Sections are considered very valuable in implementation of MG cultural program for Germany. Land MFA&A officers should be directed to prosecute vigorously activities connected with re-establishment of museums and galleries, art exhibitions, lectures, travel of German art personnel, export modern art and related undertakings which contribute to the re-education of the German people and the establishment of international cultural relations."[56]

In an official memorandum to Smyth, Standen was somewhat less direct in her criticisms, yet still very explicit.

Standen began by quoting Military Government Regulation #1 (August 18, 1947), which stated that "the reconstruction of the national German culture [had] a vital significance for the future of Germany." According to Standen, this directive was not being followed in the fine arts because OMGUS did not have a Culture Division. Furthermore, ICD and the Education and Religion Branch monopolized cultural policy without consulting MFA&A; there was no collaboration: "As things stand, the right hand does not know what the left hand is doing; the man with the 'know-how' (the MFA&A officer) has no contact with the man with the goods (the ICD officer)." Standen believed this was hurting the OMGUS reorientation agenda, because ICD's limited efforts in the fine arts were being outmatched by the French and the Soviets.[57] In her opinion, MFA&A officers had to be allowed to move beyond "personal contact and tactful advice" and encouraged to take "practical steps" to sponsor exhibitions of international, American, and German art, to organize lectures by American art scholars in Germany, and to facilitate the export of modern German art.[58] Standen did not object to ICD's involvement in procuring funds, facilities, and organizational expertise, but she wanted MFA&A personnel be in charge of art policy.

Lehamnn-Haupt and Standen succeeded in their efforts to incorporate the fine arts into the American cultural propaganda agenda. In March 1948 the Restitution Branch of the Economic Division, including MFA&A, was transferred to the Property Division of OMGUS. When the cultural restitution program ended in December 1948, the remaining MFA&A functions were transferred to the Cultural Affairs Branch of the Education and Cultural Relations Division. From then on, MFA&A dealt directly with cultural policy—its tasks included both the restitution of stolen art *and* intervention in the contemporary German art scene.[59]

Overt and Covert American Actions in the German Fine Arts

The resurgence of modern art in Germany after 1945 is often depicted as a grassroots phenomenon, an explosion of creativity after twelve years of totalitarian repression. In the American zone and sector, however, a small group of American cultural officers created the context for this revival. Starting in 1946, these officers implemented overt and covert operations to retain German artists in the American zone and sector. They fostered political and personal links between German artists and the democratic West by forming cultural associations, subsidizing artists and art prizes, and sponsoring exhibits and publications. Moreover, they used private institutions and private funds to avoid governmental scrutiny.

This small group of American cultural officers not only created an intellectual atmosphere that allowed the reemergence of modern art in post-Nazi Germany but helped to forge long-lasting intellectual, affective, and aesthetic bridges between German artists, the German public, and the democratic West. Moreover, the discreet and understated combination of overt and covert measures provided a model of intellectual warfare and cultural control that later became—greatly developed and lavishly funded—the modus operandi of the CIA in the cultural field.[1]

Prolog

Prolog, an art-appreciation group that brought together Americans interested in the German fine arts and German modern artists in Berlin, is a prototypical "private" operation in the cultural field. Formed in November 1946, Prolog was described by its organizers as a private American-German initiative aimed at helping German artists. Although the group claimed no connection to the American military government, the Americans who formed it were OMGUS officers acting within the framework of the larger reeducation agenda: MFA&A's Hellmut Lehmann-Haupt, Paul Lutzeier, chief of employee utilization of OMGUS (Berlin), and Beryl McClaskey, an OMGUS information control officer working on issues related to displaced persons and on religious matters.[2] Lutzeier and McClaskey were in contact with both Richard Howard, the head of MFA&A, and Alonzo Grace, a consultant to OMGUS on culture and education.[3] Lehmann-Haupt suggested that General Clay himself had inspired the project.[4]

Prolog was undeniably a political project. The group, McClaskey explained, created political links between Germans and Americans, "accelerating understanding . . . and encouraging democratic institutions."[5] The American members informed the German artists about cultural developments abroad, placing particular emphasis on new American art, and provided them with financial, social, and professional support. Lutzeier's and McClaskey's Berlin homes became art salons, where they entertained their German protégés, exhibited their work, and helped them sell. The artists received CARE packages and art supplies, were connected with artistic networks in the United States, and were encouraged to exhibit their work abroad. Prolog even published its own series of art books, the "Prolog catalogs." The aim was to counter SMAD's cultural initiatives and to ensure collaboration with the American political agenda once the occupation ended.[6]

Prolog's American founders were careful not to present themselves as a pressure group for modern art. The artists who participated were modern artists, many of them classified as "degenerate" by the Nazis, but their modernism was not the ostensible reason for their selection. McClaskey, in a fund-raising letter to the Smithsonian Institution, explained that Prolog was "especially interested in encouraging and supporting members of the professional and artistic community who because of their politics, race, religion or art were victims of the Nazi regime."[7] Even in 1950, when Prolog entered the mythology of the Cold War, there was no mention of the fact that it was a modern art group.

In a situation where the distinction between conqueror and conquered was very evident in daily life, Prolog was a social niche that allowed for the development of intellectual and personal relations between Americans and Germans. Prolog gatherings "were full of a special excitement and a thrilling fresh spirit of cooperation," Lehmann-Haupt wrote.[8] The German artists were treated as friends and colleagues rather than unapproachable, defeated enemies. For Friedrich Winkler, an artist in Prolog, the group symbolized the end of the isolation imposed by being German in the postwar context:

When, in the second winter after the war, you opened your hospitable homes to Germans for a cultural exchange of ideas, you appeared to us as the dove with the olive branch after the deluge that the second world war has been. We have planted this thing in the earth and it has thrived beautifully ever since.[9]

Kurt Hartmann, the publisher of the Prolog catalogs and a member of the group, recalled an experience that made him feel that Prolog was a real challenge to the intellectual and social separation of Americans and Germans:

Thus did a cup of coffee play a role in my life—a cup of coffee which you, Dr. McClaskey, offered me in [sic] December 29th 1946 in Truman Hall. This cup of coffee, handed by an American woman to a German was for me the first proof that the war was at last coming to an end. Now there would again be contacts between human beings, transcending the strictly official tone of relationships between the conquerors and the conquered. A human being would again be able to speak to other human beings. In the noisy cafeteria amongst all the loudly chattering people suddenly a new guest had come—a guest from another world. I believe, that others must have sensed the presence when this guest stood beside you and smilingly helped bridge the deep moat of misunderstanding—"Nehmen Sie noch eine Tasse Kaffee" [Help yourself to another cup of coffee].[10]

While groups such as Prolog did not erase postwar power differentials, they did contribute to establishing affective and political links between Americans and Germans.

Initially, Prolog was not an anticommunist group. Indeed, Karl Hofer, a Communist who had been appointed director of the Academy of Fine Arts in Berlin by SMAD in 1945, was an important member. It was only after the Berlin Blockade of 1948 that Prolog was incorporated into the folklore of German "resistance" against totalitarianism. In a June 1950 *HICOG Information Bulletin,* Peter F. Szluk wrote:

Just as in 1848 when Berlin's cultural leaders fought side by side with their fellow citizens for basic human rights and freedom of expression, today's artists . . . are again in the vanguard of the fight for democracy and freedom. This was evidenced, particularly, to American personnel who resided in Berlin during the days of the Blockade and were associated with the Prolog Group.[11]

The image of the artist as resistor against Soviet oppression smoothly replaced the trope of the artist as a victim of Nazism. The German artist in the American zone and sector was portrayed as the embodiment of resistence and moral courage. The West German authorities embraced this, and in 1950 West Berlin established an annual award of five thousand deutsche marks for outstanding artists, in "recognition of the roles these cultural elements play in the resistance of the people to dictatorial forces and totalitarian ideology."[12]

MFA&A and the Holocaust

Although MFA&A was never officially engaged in the promotion of modern art in Germany, Lehmann-Haupt and Howard were responsible for the first exhibit of American modern art in postwar Germany. *Ausstellung: Henry Koerner U.S.A. 1945–1947* was on view at the Haus am Waldsee, the *Kulturamt* of the Zehlendorf district in the American sector of Berlin, from March 30 to May 11, 1947.[13] Showing the work of Henry Koerner was a daring choice. It was modern art, and political art, by a Jewish Austrian American whose family had been exterminated in the Holocaust. Therefore, the exhibit ran the real risk of alienating German viewers.

Born into a Jewish family in Vienna, Henry Koerner (1915–1991) fled to Milan in 1938, after the Nazis annexed Austria, and in 1939 emigrated to New York. His parents, his brother, and other relatives, however, were deported from Vienna and murdered in Nazi extermination camps.[14] Koerner became an American citizen and worked as a commercial artist. In 1942 two of his posters won first prizes in the National War Poster Competition at the New York Museum of Modern Art. A year later he joined the Office of War Information as a graphic designer, working under the direction of Sam Rosenberg in a team that included such prominent American artists as Ben Shahn, Bernard Perlin, David Stone Martin, and Irving Miller. Koerner enlisted in the U.S. Army Corps of Engineers but was transferred to the Office of Strategic Services (OSS). He was sent first to England, then entered Germany as a member of the occupation forces, where he served as the official sketch artist at the Nuremberg Trials. He referred to this experience as "the phenomenon of a preordained situation—that I should meet the murderers of millions of Jews and my parents."[15] In 1946 he moved to Berlin to serve in the Graphics Division of OMGUS. His first one-man show was the exhibit organized at Haus am Waldsee by MFA&A.

Koerner was a magical realist, and his work was anecdotal, modern in execution, and explicitly political. He was influenced both by German artists of the Neue Sachlichkeit such as Otto Dix and George Grosz, and by his friend Ben Shahn—one of the artists ridiculed by the congressional foes of modern art. Koerner's paintings from the period 1945–1947 dealt with loss, separation, trauma, violence, and death and made visible German criminality, collaboration, and complicity. In *Der Prophet* ("The Prophet") a group of men and women are enthralled by the rants of a speaker in the foreground, while in the background a city burns, a body hangs limp, and a man covers his face to avoid seeing the disaster that is engulfing him. *Lebenspiegel* ("Vanity Fair"), which became the most widely reproduced of Koerner's works, shows a figure leaning out a window who witnesses a murder in the midst of a serene landscape. The perpetrator is stomping on the already fallen body as laborers toil, oblivious, in the adjacent field. Discussing *Lebenspiegel* with a reporter from *Time* in 1948, Koerner asked, "Who is guilty, the man who kills or those who turn their backs?"[16]

The Berlin show was intensely debated. Many American critics agreed that Koerner was an excellent young painter. *Life* reported that Koerner's work "created a sensation in Germany" and claimed that his paintings were "the best ... to date to have come out of the aftermath of the war." The *New York Times* wrote that the "American's one-man show is highly praised by Germans." And *Time* referred to Koerner as "the find of the year."[17] Others were less enthusiastic. Helen Boswell—Richard Howard's wife, who wrote a regular column for the *Art Digest*—found Koerner's choice of themes objectionable. Koerner would become a good artist, Boswell maintained, only when he "gets off his hysterical I-told-you-so path and settles down." Jo Gibbs, also writing for the *Art Digest* from Berlin, commented that Koerner's works "are concerned with the human and material wreckage of the war Germany brought on herself, and, to an extent, with the culpability and gullibility that was its cause." In his view, while technically "splendid," the artist "still has a way to go, because he is still, understandably, filled with savage bitterness."[18]

The German critics writing in the American-licensed German press refused to deal with the issue of the Holocaust, even though the killing of Koerner's family was noted in the exhibit brochure. *Heute,* a popular magazine overtly controlled by OMGUS, reported that Koerner had "the critics divided" with his "stark, almost brutal canvases." Although *Heute* reproduced the same photographs of Koerner's work that were shown in *Life,* it did not mention Nazism or the Holocaust, while the American magazine did. Edwin Redslob, the most prominent German art critic of the postwar period, took a similar rhetorical tack in his laudatory review for the Berlin newspaper *Der Tagespiegel.* Redslob wrote that Koerner had lived in Vienna and then in New York, but he did not mention why the artist had left Vienna and made no reference either to the fact that Koerner was Jewish or to the fate of the artist's family.[19] By ignoring Koerner's past, these critics were able to bypass the discussion of the Nazi genocide and the responsibility of the German people in it.

The Art Intelligence Unit of MFA&A, led by Lehmann-Haupt, designed a questionnaire to determine the impact of the paintings on German viewers. Its eleven questions were prefaced by a paragraph explaining that answering the questions was not mandatory. The form could be completed on the spot or at home, anonymously or signed, in complete sentences or in note form. These user-friendly directives suggest that MFA&A was trying to maximize the number of respondents. Although the exhibit had 1,500 visitors, only 210 (14 percent) completed and returned questionnaires. The respondents were intrigued by the content of Koerner's work, which they found unsettling, depressing, dramatic, thoughtful, melancholic, existentialist, and extremely pessimistic. Many claimed to have gone to the exhibit because they were interested in seeing modern American art but left with the impression that Koerner was not "typically American." Young Germans were the most impressed by the exhibit.[20]

Koerner's work did differ in content from other art being shown in the American zone and sector. German art in the immediate postwar period was commonly atemporal and acontextual, detached from the immediate social and political reality, and lacking a testimonial component. American art critics who visited Germany had commented on the lack of references to the disasters of war.[21] There were almost no visual references to the Nazi genocide, to the destruction of German cities, or to personal histories of terror, repression, and bereavement. Idyllic landscapes, still lifes, flowers and animals, and timid abstractions predominated.[22]

The Koerner exhibit was the only show sponsored by OMGUS that included artwork specifically concerned with recent German history. There were artists in Western Germany who dealt with social and political topics, but OMGUS never offered prizes, funding, or commissions for artists exploring the Holocaust.[23] The memory and the representation of genocide had no place in the OMGUS conception of the cultural heritage of Western Germany, and this fit comfortably with the German reticence to remember. For many years to come, West German art reflected neither the viewpoint of the perpetrator nor the experience of the victim. Following the Koerner exhibit, the Haus am Waldsee showed American children's drawings, and OMGUS never again sponsored art related to the Holocaust.

Cold War blackmail

The Cold War created a space for Germans to negotiate with their military occupiers. The competition between SMAD and OMGUS, particularly in Berlin, gave Germans leverage—they could extract some concessions by threatening to move to the other zone or sector. Both military governments had to cajole, or directly buy, the allegiance of the Germans.

On August 5, 1948, in the midst of the Berlin Blockade, a group of about sixteen hundred German artists sent a collective memo to Colonel Tom Hutton, the acting chief of the OMGUS Information Services Division in Berlin. They warned that if they did not receive support from the American authorities they would be forced to accept, however unwillingly, an offer of monetary and artistic aid channeled through the "Free Trade Union," a Soviet-controlled artist organization:

> All the artists are now in great distress on account of the currency reform. . . . In addition they have to suffer from political events, for instance the blockade. [The artists will] try to disassociate themselves from the FDGB (Free Trade Union, Russian licensed) . . . [but] the FDGB is ready and in a position . . . to put at the disposal of the artists a sum, that however is made dependent on terms endangering democratic freedom. . . . Politically they make use of the fact that the Western Allies did not do anything to assist the needy artists. The reason of our application is to ask you to put at our disposal a larger account, not connected with any political terms or links.[24]

The artists' position amounted to blackmail. Accepting Soviet economic assistance would, they claimed, have important ideological and political consequences, but they were prepared to do so unless OMGUS made a concrete counteroffer. They requested financial assistance and asked that OMGUS help create "an independent professional union to be based on democratic views and representing an unpolitical union, with no political connection to one of the Allied Powers or one of the political parties."[25] This OMGUS-financed organization would guarantee the artists' anticommunist and pro-Western commitment while preserving the appearance of artistic autonomy in West Berlin.

A massive flight of artists to the East would have been an embarrassment for OMGUS. Colonel Hutton decided that a positive response to the artists' demands was necessary to "achieve both important economic and political objectives." On August 9 Hutton summoned a group of "upper-bracket artists and art patrons of Berlin"—both American and German—to work out the details of the plan.[26] Of the ten people present at the meeting, five were associated with OMGUS or Prolog: Berryl McClaskey, Eline McKnight (American member of Prolog), Renée Sintenis (German artist associated with Prolog), Kurt Hartmann (Prolog collaborator and publisher), and John Ritter (OMGUS). The other participants included Professor Will Grohmann, the art critic for the U.S.-licensed *Die Neue Zeitung*, Karl Buchholz, the owner of a modern art gallery in Grunewald, and Hans Gaedicke, the author of the collective letter. OMGUS internal reports on the meeting refer to Gaedicke as the leader of the "Western faction" of the Artists' Union.

As a result of this meeting, OMGUS/Berlin agreed to organize and finance small sale exhibitions in studios set up in OMGUS buildings. For three to four months the artists would not have to pay rent, and after that they would be charged only a token amount. The sales would provide "temporary economic relief" for the artists, and the studios would serve as "headquarters for the artists in the American Sector."[27] In return, the artists were asked to submit confidential monthly reports to OMGUS and to create a central employment bureau so that artists would be available immediately if OMGUS required their services.

From the perspective of OMGUS, this was a political rather than an artistic operation, referred to in files as "Political Project: Graphic Artists in Berlin." Hutton regarded the American proposal as "a counter-inducement that would more than offset the FDGB offer." An "easy method of taking the graphic artists of Berlin out from under Soviet influence would help to strengthen our public position," he wrote. "Free graphic arts constitute a powerful weapon of democracy." [28]

The project was to be covert. According to Hutton's August 16 memo, "Military Government [had to] remain in the background, our sponsorship known only to a small committee which has been checked."[29] Secrecy was necessary because the agreement violated one of the cherished tenets of American propaganda—the notion of artistic freedom. This was not a trivial issue. The American military government wanted to preserve its reputation as the defender of artistic freedom, in distinction to the Third Reich and SMAD.

Yet OMGUS was buying the allegiance of artists, much as SMAD was doing. The secret nature of the Hutton agreement allowed the Americans to claim the moral high ground. In February 1949, for instance, Theodore A. Heinrich, chief of the MFA&A Section, wrote that there was "no direct official patronage of artists whatsoever"—that American involvement in the fine arts in Germany was limited to "private patronage" and "personal initiative." At the same time, he reported, the Soviets were "harnessing the arts for their propaganda value," particularly in Dresden and Berlin.[30] By keeping OMGUS's "counter-inducement" measures quiet, it was possible to criticize the Soviet strategy in the context of deepening Cold War tensions. The American press echoed the OMGUS message. In 1955 Charlotte Weidler wrote in the *Magazine of Art:*

The Russians are trying hard to win Berlin's artists over to their camp. They try to bribe them with fantastic offers, make unlimited promises and finally threaten them and their families. They invariably receive icy refusal. Berlin's artists are realistic; they cannot be bought. Although poor, often in debt, they refuse to accept promises of financial assistance by the Communists.[31]

The covert manipulation of individuals and groups in order to transform them into political assets became, a few years later, the standard American strategy in the cultural field.

Starting in the 1950s, the CIA covertly recruited, sponsored, and funded people, professional associations, and foundations to develop its Cold War agenda, and used private institutions and business as covers to finance and execute political actions in the cultural field. In some cases, the secrecy was so great that individual participants did not even know they were collaborating with American intelligence. These types of operations allowed the CIA to use the assets it considered appropriate while avoiding congressional scrutiny and, potentially, interference. For example, the opposition to modern art within the U.S. Congress led high-level American intelligence officers in Germany and elsewhere to rely on the CIA and private initiative to develop their programs of cultural diplomacy.

Free agents

In April 1948 OMGUS authorized a German tour of *Zeitgenössische Kunst and Kunstpflege in U.S.A.* (Contemporary Art and the Promotion of Art in the U.S.A.), an important exhibit of American nonobjective (abstract) art organized by Hilla von Rebay. This was a private initiative, organized with nongovernmental funds, and managed by a private institution. Unlike *Advancing American Art,* it did not involve federal funds and so was beyond the reach of the U.S. Congress.

Hilla von Rebay was an Austrian expatriate who founded and was the first director of the Museum of Non-Objective Art in New York, the forerunner of the Solomon R. Guggenheim Museum. Rebay believed that abstract art was the ultimate expression of modern art, and the art of the future. In the aftermath of World War II, she used the resources of the Guggenheim Foundation to to sponsor nonobjective art exhibits in Germany, and to send CARE packages containing food, clothing, and art supplies to artists, art historians, dealers, and critics supportive of modernism, in Germany and elsewhere in Europe. Rebay's personal connections and institutional clout served her well. Through her cousin Mimi Killian Koch, wife of art scholar Carl Koch, she established contact with a small group of painters and art dealers in Berlin, including Lotte Konnerth, Gerd Rosen, and Wolfgang Frankenstein. Thus, she established a network of German intellectuals concerned with the development of modern art in the American zone and sector, and soon became a central figure in the postwar German art scene. Many German abstract artists and art scholars have attributed the revival of modern art in Germany to Rebay.[32]

Franz Roh, a prominent postwar art critic and the leading German supporter of modern art, became Rebay's principal contact in Germany. Roh had been fired from the University of Munich by the Nazis in 1933 and was reinstated by the Americans in 1946. He came into contact with Rebay in January 1947, when he asked her for color slides for his courses. She sent him slides, and later food packages and art aid, asking in return for a short list of talented nonobjective German painters.[33] It was precisely in this way—by exchanging aid for contacts—that Rebay was able to identify a group of young German abstract artists to support.

When Rebay organized *Zeitgenössische Kunst and Kunstpflege in U.S.A.,* Roh was appointed the official spokesperson for the exhibit, and when Rebay visited Germany in 1948, Roh took her to the Moderne Galerie recently opened by Otto Stangl, a collector of German abstract art. Stangl was added to Rebay's CARE package recipient list, and Rebay asked him too for a list of nonobjective German painters. Stangl gave her the names of the painters in his monthly discussion group, and she contacted them. In June 1949 these artists—Willi Baumeister, Rupprecht Geiger, Gerhard Fietz, Willi Hempel, Theodor Werner, and Fritz Winter—formed ZEN 49, the most influential abstract art movement in southern Germany. Rebay, who "offered material and spiritual support" through the Guggenheim Foundation, was invited to become an honorary member of ZEN 49. The group held its first exhibition in the Munich Amerikahaus.[34]

Rebay's exhibit, the first showing of American abstract painting to be seen in postwar Germany, was made possible by the fact that her second cousin, the art historian Kurt Martin, the director of the Karlsruhe Staatliche Kunsthalle during the Weimar era, was working for MFA&A. It was Martin who arranged for the exhibit, which was shown in the Salon des Réalités Nouvelles in Switzerland, to travel to Germany. *Zeitgenössische Kunst and Kunstpflege in U.S.A.* opened in the Karlsruhe Staatliche Kunsthalle, then traveled to Munich, Stuttgart, Mannheim, Frankfurt, Düsseldorf, Braunschweig, Hannover, and Hamburg. It received a good deal of publicity on the radio and in newspapers, and a film was made about it and shown in movie theaters across the American zone. Lehmann-Haupt referred to the exhibit as a "valuable experiment with popular art education."[35] Not surprisingly, the United States Information Center took over the administrative side of the exhibit in 1950, sponsoring further showings in other German cities.

Financing by proxy

The first American-sponsored modern art competition in postwar Germany was also financed with private funds. The Blevins Davis Prize, organized by Stefan Munsing, the chief of MFA&A Bavaria and the head of the OMGUS Munich Collecting Point, was meant to foster the development of German modern art and to encourage young German artists. The first prize consisted of a thousand dollars and a trip to the United States, and ten runners-up also received cash prizes. The jury for the prize included Max Huggler (Switzerland), Jean Leymarie (France), and Henry Varnum Poor (U.S.), as well as prominent figures in the German art world, such as Werner Haftmann, Ewald Mataré, and Willi Baumeister. The contest was a great success, with more than thirty-seven hundred entries. One hundred seventy finalists had their works assembled in an exhibit at the Munich Collecting Point in 1949.[36]

Funding for the art prize was provided by Blevins Davis, a New York theatrical producer, millionaire, and philanthropist.[37] Davis, who described himself as a self-appointed ambassador of good will, believed that the arts established mutual understanding between nations. In the summer of 1948 Davis spent seventy thousand dollars to take the Robert Breen production of *Hamlet* to the annual Shakespeare Festival in Copenhagen, Denmark. The play was presented in association with the Cultural Relations Division of the U.S. State Department and the U.S. Air Force. After Copenhagen, the troupe toured five cities in Germany. The production was well received as a sign of American cultural vitality. Davis and Munsing met during this period, and Davis agreed to underwrite a prize aimed at rewarding young German painters and sculptors.

The Blevins Davis Prize was a cultural turning point for Western Germany. The contest integrated young German artists into the international art scene and served to further German-American cultural relations. According to Thomas Grochowiak, whose work was chosen for the exhibit, being selected was an honor not only for the financial reward but also because the prize was American. The prize thus helped to consolidate the image of the United States as a cultural leader interested in promoting the revival of German culture.[38]

The religious painter George Meistermann won the first prize with *Der Neue Adam,* and Leonhard Wuellfarth's *Almabtrieb* won the second prize. Both works were examples of the "classical modern" style favored by OMGUS cultural officials.[39] *Der Neue Adam,* heavily influenced by Picasso's *Guernica,* was a figurative religious painting about Christian redemption and rebirth. It shows a man emerging from an open hand—the new man placed by God in a new world. Franz Roh wrote that Meistermann "stands alone in that he tries to bring the new art of abstraction into the Church."[40] Like *Der Neue Adam, Almabtrieb* made no direct reference to the Nazi past or to the war. This oil, later described in a HICOG bulletin as "a pastoral scene in typical Bavarian motif," highlighted the new normalcy of Germany, totally recovered and untainted by war and criminality.[41] It was the revival of Goethe's Weimar, a sanitized German landscape uncontaminated by the Nazi rhetoric of "Blood and Soil" and devoid of crematoriums.

Christian thematics were popular in postwar art in Western Germany. Otto Dix's series on Christ (1946–1949), a complete departure from his Weimar period in terms of both style and theme, exemplifies this tendency. Groups such as the Gesellschaft für Christliche Kultur organized exhibits of religious art with some regularity. Within art theory, Hans Sedlmayr continued to champion a strong conservative Catholic trend and published two diatribes against modern art and modernism: *Verlust der Mitte* ("The Lost Center") and *Die Revolution der modernen Kunst* ("The Revolution of Modern Art").

Kultur and art

OMGUS cultural policy was not a simple exercise of Americanization by force. There is no doubt that the United States imposed American "low culture" and American political values on West German society. Yet the military government also made an effort to use German *Kultur* as a means to create a new sense of national pride that was rooted in German national heritage but devoid of political nationalism. Furthermore, the Americans placed German *Kultur* within the Western European cultural tradition. This approach coincided with that of anticommunist German intellectuals, who strove for an "Occidental" Germany, free from Nazi atavism and Soviet "Orientalism."[42]

Already in 1946, an unsigned article in the ICD *Information Control Review* stated that "the guiding images for the majority of noncommunist Germans [were] cultural and not political symbols":

> The almost religious emphasis on *Kultur* fulfills several important functions for Germans. Dwelling on the past glory of German culture restores some measure of self-respect and national pride, within a permitted context, and the hope of rebuilding Germany's prestige in the eyes of the world. It supplies an avenue of escape from the harsh economic and political realities of present-day German life. In effect, it supplies a symbol, something Germans can believe in and cling to in the midst of the general spiritual and intellectual wreckage.[43]

The importance of *Kultur* for the German public was acknowledged in 1947 with the Joint Chiefs of Staff Directive 1779. JCS 1779 warned that "there should be no forcible break in the cultural unity of Germany."[44] Indeed, cultural revival became one of the pivotal long-term goals of reeducation. The German cultural tradition and German "contributions to civilization" were to provide, according to the OMGUS "Long-Range Policy Statement for German Re-education," a source of national pride and "German self-respect."[45] In July 1947 a restricted report from the Research Branch of OMGUS/Berlin explained that German *Kultur*, supposedly immune from political deformations, had to be integrated into the rhetoric of German self-renewal and national regeneration:

> For the German who is ashamed of Hitler, it is a comfort that Johann Sebastian Bach was a German. For the German who feels humiliated by Germany's present position in the family of nations, it is a comfort that

Germany produced artists as great as those of the other nations. For the German who despairs of the mentality of his countrymen, there is the hope that "Kultur" would be the force of regeneration.[46]

In 1949 William G. Constable, the curator of paintings at the Boston Museum of Fine Arts, was appointed expert consultant in the OMGUS Education and Cultural Relations Division. His job was to define an overt and explicit fine arts policy for OMGUS. Constable was concerned with the dangers posed by German isolationism and by the persistence of extreme cultural nationalism nearly five years after the end of the war. He recommended that OMGUS organize exhibitions of American, European, and non-Western art in order to emphasize "the common cultural heritage of mankind" and widen the outlook of the German population. American museums, universities, and cultural centers had to become involved in Germany's cultural revival, American art scholars had to travel to Germany, and the Amerikahäuser had to be transformed into reference centers on American art and culture.[47]

At the same time, Constable argued that the Americans should build on the non-Nazi German cultural tradition. He believed that the OMGUS cultural policy should emphasize those aspects of German culture that fit within the American conception of a new Germany:

[OMGUS must] select and emphasize those elements in the German tradition which are most in consonance with American aims. I have found, for example, in trying to drive a point home that to quote the example of (say) Duerer, Bach or Goethe, or even the Behest, changes the whole atmosphere of discussion.... [Democratic and liberal ideas] can be profitably grafted on to the German tradition as a logical means of strengthening its best elements.... The German people have a long and vigorous cultural tradition of their own. To attempt to substitute for it an American or any other tradition is impossible, and will only stimulate refusal to learn, and lead to ultimate rejection.[48]

In Constable's view, *Kultur*—and American respect for *Kultur*, evinced in such programs as exhibits of German Old Masters in the Amerikahäuser and at OMGUS central collecting points—could be a source of pride and national identity for Germans seeking a cultural patrimony free from the scourge of National Socialism.

The reemergence of modern art in West Germany was fraught with difficulties. Modern artists were confronted by the Nazi legacy of cultural antimodernism as well as the aggressive antimodernist campaign waged by SMAD. OMGUS cultural officers, too, had to be cautious in promoting modern art because of the antimodernist ideology prevailing in the U.S. Congress. Yet by 1949 the development of an official fine arts policy aimed at selectively recuperating German cultural traditions and fostering a new, nonnationalist German art scene had become an OMGUS objective. Unlike the Nazis and the Soviets, OMGUS had minimal official institutional involvement, and virtually no overt financial involvement, in the West German fine arts. It was only through covert cultural policies and private financing—bypassing the scrutiny of the U.S. Congress—that it was possible to advance a timid promodernist agenda in post-Nazi Germany and to establish links between West German artists and art critics and American artists and art institutions.

Five

Iconoclasm and Censorship

Visual censorship always accompanies radical regime change. Major religious and political upheavals—the Reformation, the French Revolution, the Bolshevik Revolution—have invariably been accompanied by violent iconoclasm. The destruction of the defeated enemy's icons figures prominently among the prerogatives of the victors.[1] In the twenty-first century, the Taliban destroyed fifth-century Buddhist monuments in Afghanistan, and American occupation forces dismantled statues of Saddam Hussein in Iraq.

By 1945 the U.S. government had extensive experience controlling the visual landscape for purposes of education, indoctrination, and propaganda. During World War I, the American Commission on Public Information (popularly known as the Creel Committee) censored images of the war, not allowing the publication of photographs of dead American soldiers. The Propaganda Section (or Psychological Section) of G-2—U.S. Army intelligence—was also involved in visual censorship during World War I.[2] In the 1930s American films were shaped and constrained by a complex web of censors—the Motion Picture Producers and Distributors of America, the Catholic Legion of Decency, the Production Code Administration, the National Board of Review, and powerful state censorship boards in New York, Pennsylvania, Maryland, Kansas, Virginia, and Ohio.[3] After Pearl Harbor, the Roosevelt administration set up the U.S. Office of Censorship, activated the War Department's Bureau of Public Relations, and created the Office of War Information (OWI), the wartime agency in charge of domestic propaganda and censorship.[4]

In occupied Germany, visual censorship became a central part of psychological warfare. From the start, the U.S. Army and OMGUS engaged in iconoclasm, removing Nazi images and symbols from the visual sphere and monitoring the German media to prevent their resurgence. With the emergence of the Cold War, OMGUS narrowed the parameters of what was permissible and began to censor images that, although clearly not pro-Nazi, challenged the American agenda in Germany. This shift is exemplified by the response of the Information Control Division (ICD) to the satirical journal the *Ulenspiegel* when it veered toward explicit anti-Americanism in 1947.

The iconoclastic campaign

On February 12, 1945, President Franklin Roosevelt, Prime Minister Winston Churchill, and the General Secretary of the Central Committee of the Communist Party of the Soviet Union, Joseph Stalin, signed the Joint Communiqué in Yalta. This document, intended as the Allied blueprint for the occupation of Germany, ordered the removal of "all Nazi and militarist influences from public office and from the cultural and economic life of the German people."[5] In May 1946 the Allied Control Council codified the denazification of the public domain through Control Council Directive No. 30:

The planning, designing, erection, installation, posting or other display of any monument, memorial, poster, statute, edifice, street or highway name marker, emblem, tablet, or insignia which tends to preserve and keep alive the German military tradition, to revive militarism or to commemorate the Nazi Party, or which is of such a nature as to glorify incidents of war ... will be prohibited and declared illegal. ... Every existing monument, poster, statue, edifice, street or highway name marker, emblem, tablet, or insignia ... must be completely destroyed and liquidated.[6]

The Allied occupation forces were instructed to purge the German public sphere of cultural objects produced or used by the Nazis to disseminate their nationalist and militarist ideology.

The American iconoclastic campaign followed both the letter and the spirit of the Joint Communiqué. First the U.S. Army, and then OMGUS, changed street names and destroyed Nazi icons in public spaces. MFA&A officers ordered the removal of the *Hoheitszeichen* (Nazi eagles) from party buildings, post offices, schools, and residences, eliminated plaques commemorating Hitler, and eradicated swastikas. Public buildings, airports, highways, and other remnants of the Third Reich that had ongoing practical uses underwent physical change and were cleansed of Nazi symbols.[7]

The Americans wanted to create a new German visual culture devoid of icons of militarism and Nazism. Joint Chiefs of Staff Directive 1067, issued May 15, 1945, ordered that "all archives, monuments and museums of Nazi inception ... devoted to the perpetuation of German militarism" be placed under the control of the U.S. Army.[8] In July 1945 the U.S. Army issued Law No. 52 prohibiting commerce in work of arts or cultural materials. Law No. 52 did not differentiate between art by traditional masters, Nazi art, and military art; "all works of art were to remain in limbo awaiting Military Government instructions as to disposition."[9] On July 24, 1945, the U.S. Army released Military Government Regulation Title 18-401.5 to regulate the implementation of Law No. 52. Title 18 ordered that "all collections of works of art relating or dedicated to the perpetuation of German Militarism or Nazism will be closed permanently and taken into custody."[10] Immediately after the cessation of hostilities, art museums and collections were closed to the public in order to avoid the exhibition of such art.

Captain Gordon W. Gilkey, of the Army Air Corps (Intelligence), was made responsible for locating and confiscating Nazi war art. Gilkey, as "coordinator of information about cultural materials found in the U.S. Zone of Occupation," was assigned to the Office of the Chief of the Historical Division, U.S. Forces, European Theater, on May 9, 1946. His task was not easy. The Nazis had hidden their war art to protect it from the Allied bombings and save it for posterity. Much of the Führer's personal collection, *Kunst der Front,* was hidden in the salt mines of Bad Aussee, in the basement of the Schloss Oberfrauenau, and in a dance hall in St. Agatha, Austria. The personal collections of Himmler, Bormann, and other Nazi leaders were stored in the *Führerbau,* in Munich's Koningplatz, and Himmler's touring show of SS art was stored in the Befreiungshalle, Kelheim. Several crates containing rolled-up war paintings, camouflaged to resemble stage curtains or hidden behind framed nudes and pastoral landscapes, were found in the cellars of the Haus der Deutschen Kunst.[11] Gilkey spent over seven months searching for and seizing Nazi art, and three additional months cataloging the confiscated collections.

Gilkey classified as Nazi or militaristic a total of 8,722 objects. The material included Nazi emblems, war paintings, portraits of or by party dignitaries, and artwork by prominent Nazi artists. On March 20, 1947, the confiscated art was shipped to the United States, to be stored in military deposits in Washington, D.C.[12] The items that remained in Germany were later destroyed as being of "no historical and art historical value." Gilkey understood that "German art [had] became a tool to spread the manure of Nazism and Nazi directed German militarism." He thought that Nazi war art, if left in Germany, "had the potential of inducing nostalgia among the prevalently pro-Nazi German population, capable of prompting memories that would rekindle old passions."[13]

Iconoclasm was accompanied by a policy of selective preservation.

Title 18 stipulated the need to protect German cultural property not associated with Nazism. MFA&A was ordered "to protect, preserve and control all cultural structures, objects and archives in the U.S. Zone" in order to avoid the "loss to the cultural heritage." OMGUS cultural officers determined which elements of German culture were worth preserving and which should be removed.[14] The American definition of a desirable and acceptable German culture, however, changed with time. By 1947 the limits of tolerance were redefined by changes in American foreign policy.

Creating and enforcing boundaries

Immediately after they occupied Germany, the Americans became concerned with encouraging the development of a new German press free of Nazi and militaristic tendencies. OMGUS aspired to create a new kind of independent German press, not controlled by the state, by the political parties, or by the church. To this end, it instituted a system of indirect control: ICD officers selected and licensed Germans to run newspapers and journals in the American zone and sector, exerted postproduction censorship, and regulated circulation by determining the paper supply for each publication. The German editors were free to run their operations, but there was always the possibility of postproduction reprimands that could lead to the removal of licenses.

The ICD personnel involved in the early American press policy had a variegated ideological profile. As in the OSS, many ICD officers in 1945–1946 were New Dealers, intellectuals, Jews, and leftists. Those in Berlin were typically young or early middle-aged intellectuals, enthusiastic about helping to build a democratic and pluralist society from the ashes of Nazism. And the majority were German emigrés.[15]

The young, noncareer officers in ICD were imbued by the spirit of Roosevelt's New Deal and social democracy. For them, the communist and socialist Germans who had fought against Franco in Spain, and those who had resisted Nazism at home were the natural allies of the United States, the true antifascists. It was unthinkable to build a democratic and antimilitaristic Germany without their participation and perspective. In 1945, then, OMGUS press officers welcomed the collaboration of the German left without reservation.

At first, ICD officers were primarily concerned with preventing Nazis from participating in the German media. OMGUS prohibited the publication of overtly pro-Nazi material and excluded journalists who had been deeply involved with the Nazi regime. Licensing for the press involved a thorough background check that was longer than the standard *Fragebogen* (political questionnaire) used by OMGUS intelligence in the American zone and sector.[16] Although the exclusion process was imperfect, the German intellectuals and journalists licensed by ICD generally had anti-Nazi credentials.

Of course, there was a tension between the professed American aim of encouraging a free press and the reality of occupation. The military government was aware of the basic contradiction, as a 1948 OMGUS report shows:

> The press officers were primarily concerned with preventing former Nazi journalists from participating in the new, democratic German press. In 1945, when the occupation began, it was a major policy of the American Military Government to guarantee to the German population an independent and free press. MG [Military Government] envisaged a press which would be free of any form of governmental domination. Yet, ironically, MG itself in 1945 found it necessary to exercise certain temporary controls. Many of the newspaper plants were in the hands of Nazis. The publishers, editors and personnel of the newspapers were the same persons who had been carrying out the policies of Goebbel's Propaganda Ministry. So MG set up a licensing system to place the newspapers in the hands of editors dedicated to giving the German people unbiased news coverage.[17]

This statement, however, is disingenuous. From the first day of the occupation, the Americans exerted total control over the German media in their zone and sector. Their objective was to radically reform the German press and catalyze the emergence of newspapers and journals that mirrored the American model. If the German press followed the American prescription, it could thrive. Editors and journalists who pursued an alternative agenda would be reprimanded or have their licenses revoked.

Nonetheless, during the first year of occupation, the distribution of licences showed a tolerance for political diversity and the development of a variegated political discourse. While enforcing a rigid ban on the diffusion of Nazi, nationalist, and militaristic messages, ICD had, by mid-1946, licensed seventy-three Germans in the press—twenty-nine Social Democrats (SPD), seventeen Christian Democrats or Christian Socialists (CDU/CSU), twenty-two with no party affiliation, and five Communists (KPD).[18]

The Ulenspiegel

In December 1945, less than six months after the unconditional surrender of the Third Reich and the military occupation of Germany, two anti-Nazi German intellectuals, Herbert Sandberg and Günther Weisenborn, launched the bimonthly satirical journal *Ulenspiegel: Literatur, Kunst, und Satire* in the American sector of Berlin. The *Ulenspiegel* published poems, short stories, and cultural criticism, but its main political text was visual, entailing caricatures, cartoons, and social/political allegories created by hundreds of German painters and graphic artists. The *Ulenspiegel* was a showcase of postwar German modern art. In its pages, artists such as Eduard Braun, Georg Grosz, Karl Holtz, Heinrich Kilger, Gerhard Kreische, Frans Masereel, Oscar Nerlinger, Albert Schäfer-Ast, and Rudolf Schlichter continued the tradition of prewar German modernism, rendering stunning examples of expressionism and surrealism to express political and social critique.[19]

Herbert Sandberg (1908–1991), the journal's art editor, was born in Poland into a Jewish family. He moved to Berlin in the early 1920s, became a member of the Assoziation revolutionärer Bildender Künstler (Association of Revolutionary Artists), and in 1931 joined the German Communist Party. In 1933 he went underground but acquired public notoriety when he climbed on the roof of KaDeWe (Kaufhaus des Westens), the largest department store in Berlin, and dispersed fliers printed with the slogan *Hitler heisst Krieg* ("Hitler means war"). Sandberg was jailed in 1935 and in 1938 transferred to Buchenwald, where he worked in the labor gangs forced to build the camp. There he was initially registered as political prisoner number 3491, then reclassified as Jewish political prisoner number 7090. After ten years of horror, he was liberated by the U.S. Army in 1945. In Sandberg's words, *Ich schien noch einmal geboren zu sein* ("I seemed to be born once again").[20]

Sandberg conceived the idea of a satirical political journal while in Buchenwald, a dream that became possible in postwar Berlin. Sandberg thought that a journal of art and satire could confront readers with the harsh realities of the Nazi past and reacquaint them with international cultural developments. His model was *Simplicissimus,* an antiestablishment journal of political satire and caricature that started in 1896 and was banned by the Nazis for being critical of their regime and ideology. *Simplicissimus* relied heavily on visual images and had impressive modern graphics. Sandberg believed that a similar journal would be important in the postwar period.[21]

In June 1945 Emil Carlebach, one of the American-licensed editors of the *Frankfurter Rundschau,* introduced Sandberg to American ICD officers. Carlebach had been one of the Communist Party leaders in Buchenwald, where he met Sandberg.[22] In the fluid months of the early postwar period, ICD did not hesitate in licensing German Communists, and Sandberg was offered an American license to publish the newspaper *Der Tagesspiegel.* Sandberg rejected the offer, and instead proposed the creation of a satirical journal, the *Ulenspiegel.* His suggestion was accepted. Sandberg, after consulting with Johannes R. Becher, the principal Communist cultural operative in Berlin and the organizer of the Kulturbund, took the American license.[23]

Every American-licensed publication had at least two editors, in keeping with the aim of intellectual diversity and ideological plurality. According to Sandberg, he proposed three possible colicensees, Horst Lommer, Paul Rilla, and Günther Weisenborn. Peter de Mendelssohn, the Berlin-born, British-nationalized ICD officer in charge of the German press in Berlin, selected Weisenborn, a Social Democrat. Sandberg recalled that Weisenborn, who was the journal's main contact with ICD, told the Americans that their intention was to make the *Ulenspiegel* a truly German journal, not a propaganda outlet.[24]

Günther Weisenborn (1902–1969), the *Ulenspiegel*'s literary editor, was born in the Rhineland into a middle-class Jewish family. In 1928 he moved to Berlin, where he became a friend and collaborator of Bertolt Brecht. With the advent of Nazism, his books were banned and later burned. In 1936 he went to New York and worked as a journalist, but a year later he returned to Germany to join the anti-Nazi resistance. He was arrested in 1942 and spent three harrowing years in Gestapo prisons. Liberated by the Red Army in 1945, Weisenborn became one of the leading intellectuals in postwar Berlin. He presided over the Schutzverband deutscher Schriftsteller (Association for the Defense of German Writers), and was the creative director of the Hebbel Theater in the American sector of Berlin. In August 1945 he staged Bertolt Brecht's *Dreigroschenoper* ("The Three-Penny Opera"). The following year, Weisenborn wrote his most important play, *Die Illegalen,* in which he stressed the courage of the anti-Nazi resistance. In fact, he was one of the first German intellectuals to discuss publicly the existence of a German resistance, at a time when neither the Americans nor the Soviets were particularly interested in the subject. Weisenborn was an organizer of the First Congress of German Writers in 1947, one of the most important cultural and political events in early postwar Berlin. In 1948 he was one of the coordinators of the famous debate in which Jean-Paul Sartre publicly defended his play *Les Mouches* against vitriolic attacks from SMAD and the German Communists.[25] Weisenborn represented anti-Nazi, noncommunist German intellectuals who wanted a united, neutral, and democratic Germany equally independent from Washington and from Moscow.

Sandberg and Weisenborn had had radically different life experiences, which fostered distinct perceptions of the reality around them. Sandberg was twenty-three years old when he joined the German Communist Party (KPD) and thirty when he was sent to Buchenwald. His formative years were spent in Nazi prisons and camps. In Buchenwald he had extensive contact with German Communists, and after liberation he established close relations with Walter Ulbricht and other members of the Stalinist hard core who had survived the purges during their exile in Moscow. Sandberg's worldview was narrower, more schematic, and more dogmatic than Weisenborn's. He did not publicly question the Stalinist take-over of the Revolution, nor the Stalinization of the Socialist Unity Party (SED), and he did not seem to notice, or resent, the increasing harshness of SMAD and the totalitarian structure that the Soviets and the German Communists were imposing in the East. Weisenborn, on the other hand, had lived in Argentina and in the United States, where he had met people with different backgrounds, experiences, and dilemmas. He was a friend of Brecht, had close relations with European intellectuals, and had an international and cosmopolitan outlook.[26] In 1945, however, Sandberg and Weisenborn shared a general commitment to combatting the reemergence of Nazism and creating an independent, democratic, and unaligned Germany. The fact that Sandberg, a Communist, and Weisenborn, a Social Democrat, were licensed to edit a journal in the American sector of Berlin illustrates the openness of the early American press policy in occupied Germany.

The Ulenspiegel, *1945–1946*
Under the joint editorship of Sandberg and Weisenborn, the *Ulenspiegel* became a resounding success. The journal's circulation jumped from 50,000 in 1945 to 130,000 by 1948. The *Ulenspiegel* reproduced thousands of German and foreign works of art and reintroduced the German public to "degenerate" artists who had been banned by the Nazis, such as Marc Chagall, George Grosz, Karl Hofer, Käthe Kollwitz, and Pablo Picasso. Moreover, the *Ulenspiegel* used satire to expose the multiple contradictions facing postwar Germany. Its editorial headquarters became a meeting place where anti-Nazi intellectuals from all over Berlin—Communists,

Social Democrats, and Liberals—gathered to discuss politics, history, and culture and to ponder the future of Germany. There, cultural officers of the occupying powers mixed with prominent Germans of the era, among them Erwin Redslob, Karl Linfert, Alfred Döblin, Hermann Henselmann, Karl Hofer, Oskar Nerlinger, Heinrich Ehmsen, and Max Pechstein. Jean-Paul Sartre was also a visitor.

During its first year, the *Ulenspiegel* focused on Nazism and the issue of German complicity. Nazism had not been a natural catastrophe, the journal insisted, but rather a criminal political system supported by millions of Germans.[27] The *Ulenspiegel* blamed the Nazis, not the Allies, for the destruction of Germany. In the first issue, Ismar Kallweit's photomontage "Architecture of the Third Reich" shows Hitler wrecking a city with a devastating gust from the sky. In Rudolf Schlichter's "Assault on German Culture," the combined forces of German militarism, imperialism, and Nazism shatter German culture. Heinrich Kilger's "The Very Small and the Voracious Moloch" depicts Hitler as a gigantic robot, operated and maintained by a legion of tiny Germans. The figures busily oil the robot, polish its surface, whisper encouragement to it, and feed the engine, while it spouts out corpses.[28]

The *Ulenspiegel* also exposed the duplicity and opportunism of many Germans in the postwar period, and expressed skepticism about the American denazification program. Sandberg's January 1946 drawing "We Have Always Been Anti-fascists," a multipanel caricature showing how Nazis were surviving—and prospering—during the occupation, exemplifies this theme. In another Sandberg drawing from the same period, a Nazi Party member takes off a shirt covered in swastikas and submerges his head in a bucket. The bucket holds a *Fragebogen* with inane questions, all answered in the negative. The Nazi emerges clean, confident, and radiant, wearing a fancy new suit and smoking a cigarette, the symbol of wealth in postwar Germany, where tobacco was the most valuable currency.[29]

At the same time, the *Ulenspiegel* introduced an emblematic figure, Michel, to portray the "common German." Michel was a weak and astonished character, a survivor trapped in a world being reconstructed by the Allies but with no say in the process. Sandberg's January 1946 caricature "The Three Holy Kings," shows a crippled Michel looking at the three magi with a puzzled expression. The first, representing the Communist Party, holds a toy plough labeled "Land Reform." The second, the Social Democrats, carries a flag with the inscription "Social Reform." The last, the Christian Democrats, holds a priest grasping a school on his lap. A bewildered Michel wonders for whom he should vote. In this illustration, Sandberg reflects on the disjuncture between the real needs of an impoverished, crippled, and dysfunctional Germany, and the platforms of the main postwar political parties. Similarly, in Sandberg's August 1946 drawing "The German Buildings," two groups of construction workers are busy on opposite corners of the same project rather than collaborating to build a single structure. The cover of the November 1946 issue is most telling: two identical Michels stand in identical ruined landscapes separated by a wall. The sky above one is the American flag; above the over is the Soviet flag. Each looks longingly at the other's sky.[30] It is evident that at least the German intellectuals gathered around the *Ulenspiegel* had an astute perception of the political situation in Berlin.

The Michel icon—wounded, mute, and confused—facilitated a transition to a rhetoric of national renewal. As early as 1946, the *Ulenspiegel*'s visual text presented two subsets of Germans, the "bad Germans" tainted by Nazism, militarism, industrialism, corruption, and wealth, and the "good Germans," innocent, powerless, confused, and poor. Gerhard Kreische's December 1945 drawing "Christmas" shows three Wehrmacht veterans surrounded by urban ruins. Cold, famished and in rags, they contemplate a spindly Christmas tree labeled "Peace." Rather than use the tree for wood, the veterans watch it with reverence and hope. An April 1946 drawing by Oscar Nerlinger also looks to the future: a couple sits among tree stumps in an empty park, facing a devastated skyline; the caption, by contrast, reassures the reader that it will be warmer soon and the Berlin *Tiergarten* will again turn green. In a March caricature by Sandberg, happy couples stroll along a sunlit road that leads to the UN, while a little Michel tries to shovel away a mountain of snow that blocks his access. Near the mound of snow—labeled "Guilt"—a snowman wearing a military cap and carrying a sword slowly melts under the spring sun. The caption reads, "Friend, we need to find the way through!"[31] For Sandberg, atonement was possible through the eradication of German militarism. The "good Germans," however, could not realize their dreams because their homeland was divided. The only alternative was the construction of a new democratic Germany following its own initiatives and policies.

The *Ulenspiegel*'s perspective on German recovery in the post-Nazi era corresponds to what is known as the "Third Way," an assertion of the possibility of constructing a new, unified Germany based on a democratic and humanist socialism, but only after a thorough denazification process. This unified Germany, Third Way intellectuals believed, would be able to escape the politics of international polarization, avoiding both "Sovietization" and "Americanization" and instead finding its place in an independent, pan-

European context. Between 1945 and 1948 a wide constellation of left-wing intellectuals—left liberals, left Christians, socialists, social democrats, and communists—pursued this idea, collaborating in political and cultural journals such as *Ost und West* and *Aufbau* (in the Soviet zone) and *Frankfurter Hefte, Nordwestdeutsche Hefte, Der Ruf,* and *Die Wandlung* (in the western zones).

The Cold War, OMGUS press policy, and the Ulenspiegel

Despite the optimism of the *Ulenspiegel* and Third Way intellectuals, the real possibility of an independent and united postwar Germany was rapidly vanishing. Nineteen forty-six was a pivotal year in German and international politics. Winston Churchill, in his March 5 speech at Westminster College in Fulton, Missouri, announced the end of the wartime alliance with the USSR. Churchill spoke of an "iron curtain" falling across Europe, making explicit the confrontational atmosphere that would characterize the Cold War. On July 11 U.S. secretary of state James F. Byrnes, speaking at the Council of Foreign Ministers meeting in Paris, invited all the occupying powers in Germany to merge the economies of their zones. Only the British accepted, and on July 30 the United States and the UK agreed to merge their zones, creating Bizonia. On September 6, speaking in Stuttgart, Byrnes announced a new American policy in Germany, explicitly calling for economic reconstruction and the reintegration of the country into the community of nations.

These events reshaped the OMGUS press policy. The case of the *Neue Zeitung* is illustrative. The newspaper was launched by OMGUS on October 18, 1945, with two goals—to provide a positive image of the United States and to serve as the model of a democratic newspaper. Its first editor was Major Hans Habe (János Bekéssy), a Hungarian Jewish journalist who had emigrated from Vienna to the United States in 1940. In 1942 Habe joined the U.S. Army, and as a military intelligence officer trained foreign emigrés for psychological warfare. To help him at the *Neue Zeitung,* he selected a group of German Jewish emigrés who shared a common cultural background and a commitment to a new democratic Germany. In its first year, the newspaper devoted a great deal of space to discussions of German culture.

In March 1946 the Americans forced the *Neue Zeitung* to change its editorial policy. The night of Churchill's appearance in Fulton, General Eisenhower's headquarters instructed Habe to devote the entire front page of the next day's paper to the speech. Although Habe agreed with Churchill's declarations, he refused to accept this type of order and resigned. He was succeeded by Hans Wallenberg, a German emigré journalist. Under Wallenberg, the *Neue Zeitung* became a mouthpiece for OMGUS. Wallenberg kept the emphasis on German *Kultur* but pointedly attacked SMAD's *Tägliche Rundschau*. By late 1946, both the Americans and the Soviets were using the German press in their respective zones and sectors as political weapons.[32]

In 1947 the Soviet Union consolidated its hold over Eastern Europe and initiated a political, cultural, and ideological offensive against the West. President Harry Truman responded with Executive Order 9835, signed on March 25, which authorized intelligence agencies to investigate the beliefs, political allegiances, and personal lives of federal employees in the United States and abroad. On May 12 Truman asked Congress for four hundred million dollars in economic and military assistance for Greece and Turkey, declaring the beginning of a global ideological war against communism. A month after George Kennan, the American chargé d'affaires in Moscow, sent his "Long Telegram," recommending a policy of containment of the Soviet Union, this became the official doctrine of the United States. In July secretary of state George C. Marshall announced the American plan to aid the economic recovery of Europe (including Germany) and the Soviet Union. The USSR rejected the Marshall Plan, denounced it as a ploy to further the aims of American imperialism, and forced its first European satellites, Poland and Czechoslovakia, to do the same. Stalin moved to establish absolute control over communist parties all over the world, subordinating their interests and their policies—including their cultural policies—to the Soviet agenda.

In Germany, both SMAD and OMGUS radicalized their cultural policy. SMAD abandoned all pretense of ideological neutrality, announcing its intention to create a new German culture inspired by Marxist-Leninist ideology and modeled on the USSR. The Soviet-controlled press in Germany (*Tägliche Rundschau*, the *Berliner Zeitung*, and *Neues Deutschland*) engaged in direct attacks against the United States. On the American side, most publications that did not follow the OMGUS anticommunist directives were either terminated or had their editors replaced by early 1947. Colonel Gordon E. Textor, a West Point graduate with an engineering degree, became head of ICD in 1947. That June he claimed that former membership in the German Communist Party (KPD) was not grounds for dismissal from a position in the German press. A Communist could keep his job in the American-licensed press if his political views were not reflected in his publication. From that point on, however, ICD heightened its surveillance of newspapers and journals with communist editors.[33]

Editors deemed unreliable were removed and replaced by a new breed of Cold War journalists. The German-Jewish emigrés who ran the *Neue Zeitung,* for instance, were accused of devoting too much attention to German cultural matters while disregarding the American anticommunist agenda. They were dismissed and replaced by a series of increasingly belligerent anticommunists.[34] *Der Ruf,* a journal originally created as part of an American program to reeducate German prisoners of war, was shut down. The journal had links to the European communist and noncommunist left, and ICD did not like its "nihilistic" and nationalist edge. Even though *Der Ruf* was popular and SMAD had denounced it as an anticommunist publication, ICD first reduced its paper allotment and then, in June 1947, revoked its license.[35] In August 1947, Emil Carlebach, Sandberg's original contact with OMGUS, lost his license for the *Frankfurter Rundschau.* Textor argued that Carlebach "possesses political beliefs and traits of character of such nature as to render him unsuitable as one of the leaders of public opinion and of the democratic free press." Carlebach was not the only victim of Textor's ICD—two Communists who were to edit a paper in Bremen were dismissed before they could start.[36]

In October 1947 General Lucius D. Clay launched Operation Talk Back, a counterpropaganda measure designed to use the German media to combat Soviet anti-American propaganda in Germany. This raised the ante. Up to then, the State Department Office of Information and Cultural Exchanges had avoided a full-fledged anti-Soviet propaganda offensive in Western Europe. Now, however, ICD expected the German editors in its zone and sector to actively defend the United States and attack the Soviet Union and SMAD. Censorship, previously restricted to expressions of Nazi ideology, now was extended to communism. A strict anticommunist line was imposed on the American-licensed German press, equivalent and complementary to the SMAD line that prevailed in the Soviet zone and sector.

The *Ulenspiegel* did not follow the new American policy, nor did the journal's visual narrative coincide with the OMGUS conception of American-German relations. In 1947 the journal continued its relentless denunciation of the complicity of German industrialists with Nazism and militarism, while portraying Germany as a victim, an impotent spectator of its own history, being bullied by foreign aggressors. The *Ulenspiegel* images questioned the role of America as a true liberator from the Nazi past, raising doubts about the American commitment to developing German democracy. The allegation was that an unrepentant Germany was being transformed into a U.S. lackey and that the American-sponsored political system that had emerged in Western Germany was corrupt and ideologically tainted by Nazism.

The anti-American slant grew progressively stronger. Americans were alternatively depicted as villains or fools—either partners, profiteers, and collaborators of the Third Reich, or simpletons unable to understand the depth of Nazi ideology in postwar German society. Throughout 1947 the journal denounced the increasing power of former Nazis, pointing out how the industrial barons of the Third Reich had transformed themselves—under American aegis—into the new Western German industrial elite. Sandberg's July caricature *"Wiedergutmachung"* describes the economic fates of a concentration camp inmate and his guard. The Nazi criminal is shown as prosperous and happy in 1947, while the former inmate—exultant with joy after his liberation in 1945—watches in desperation as his furniture is impounded by creditors. At the same time, the *Ulenspiegel* criticized the superficiality and inconsistency of the American denazification process. Jean Effel's December 1947 drawing "With Other Eyes" shows a flirtatious American military policeman asking a German farm girl with long braids if there are still Nazi influences in the area. All the farm animals around her have Hitler's face, as does the moon.[37]

The Soviets and the Americans were conceptualized differently in the *Ulenspiegel*. The journal satirized the emergent Western German political parties as servants of the occupiers but did not comment on the fact that the Socialist Unity Party (SED), known in Germany as the "Russian Party," was controlled directly by SMAD and the Kremlin. The journal's visual text exposed the way in which Western Germany was being shaped by the United States, but it did not denounce how the USSR was shaping Eastern Germany. Sandberg's 1947 collage "Freedom and Equality," in which he denounced SMAD's campaign against modernism, was a rare critical reference to the politics of the Soviet Union, SMAD, and the SED.[38]

By the end of 1947, the theme of war and peace dominated the *Ulenspiegel*'s images. The "good" Germans, they implied, wanted peace and unity (the central trope of Soviet propaganda at the time). The blame for instability and division was sometimes placed on both the Soviets and the Americans, at other times on the Americans alone. Sandberg's April watercolor "Conversation at the Border" shows the angel of peace asking a Soviet soldier permission to enter Moscow; the cheerful soldier replies that she needs permission from the "others"—the Western powers. In a September cover titled "Fifty-Fifty," Sandberg depicted the torso of Europe (a middle-aged woman) sitting on a bench next to the torso of Germany (a young girl). Both watch with apprehension as their lower bodies walk away, while Uncle Sam and a Russian officer look on from opposite sides of the page. The caption reads, "Is this the way the world is going?" Karl Holtz's watercolor "The Right Way," in the October issue, shows a baffled Michel standing at a fork in the road. There are two signposts, one pointing to the left and one to the right, each marked with an arrow and the notice "The way to democracy." The caption reads, "The left way seems to be the right way after all. Until now we have followed the path of the right and it has only brought suffering and blood."[39]

The Ulenspiegel *in 1948*

In 1948 ICD and the *Ulenspiegel* continued along diverging paths. In February ICD began its Vigorous Information Program (VIP), directed toward protecting "democratic principles [from] all forms of totalitarian and police-state influences, whether they be from former Nazi and fascist or from Communist doctrines."[40] All media under ICD control were to diffuse positive information about American democracy, American freedom, and the American "way of life," while criticizing Communism, the USSR, and the Soviet system. The main instruments for this campaign to sell the American model of society were the *Neue Zeitung,* Radio in the American Sector (RIAS), the *Berliner Blatt,* and the U.S. Information Centers.

The Soviet and the American propaganda campaigns attempted to introduce different modalities of nationalism. The Soviet strategy in Germany was to "give the appearance of local political power, under the banner of the national independence and anti-fascism."[41] As in the non-Russian Soviet republics, the goal was to imprint a national identity devoid of sovereignty, endowed with foundational myths and a historiography made in Moscow. East Germany would exhibit all the conventional cultural attributes of a nation-state, but the real political, economic, and military power would reside with the USSR. The Americans, on the other hand, shaped a new collective identity for West Germany that was based on pride in being democratic, liberal, tolerant, and capitalist. This new identity was stripped of the conventional attributes of national identity—chauvinistic pride, militarism, the conception of fatherland. Being a West German meant belonging to a liberal and capitalist Germany linked with Western Europe and the United States.

During its last year in West Berlin, the *Ulenspiegel* advocated a new German nationalism, defined by the rejection of both Nazism and the United States. This "progressive nationalism," purely cultural and devoid of militaristic and expansionist connotations, could only emerge within the context of a new Germany unified under "socialism." Even after the Soviet coup in Czechoslovakia in February 1948, the *Ulenspiegel* continued to associate a progressive German nationalism with socialism. This conceptualization of German nationalism fit within a wider political stance, as the slogans of the Third Way became absorbed into East German political rhetoric.

By May 1948 OMGUS could no longer tolerate the *Ulenspiegel*. ICD officers began looking for a new editor and reduced the journal's paper supply. The number of copies that could be printed fell first to fifty thousand, then to twenty-five thousand. The decision to reduce the *Ulenspiegel*'s circulation was political rather than economic, as an internal ICD memo shows: "Orders have been issued to reduce by one-half the paper allocation given to the magazine *Ulenspiegel* as a preliminary step toward inducing a change in editorial orientation of this publication or its replacement with a more effective medium."[42] ICD insisted that it was ready to restore the journal's paper allotment if the *Ulenspiegel* changed editorial policy to comply with its demands.

ICD officers misunderstood, however, why the *Ulenspiegel* had developed its critical narrative. They believed that the editors were trying to ingratiate themselves to SMAD in order to ensure their position if the United States pulled out of West Berlin:

Various indications point toward an attempt on the part of our licensees to avoid actions or statements which might prejudice them in the eyes of the Soviet occupation authorities. This trend is undoubtedly an inevitable product of the uncertainty surrounding, in the eyes of the Germans, the permanence of the American position in Berlin. It is being countered decisively through such measures as reduction in paper allocations to those publishers who show faintness of heart.[43]

ICD officers apparently could not conceive of, or accept, that West Berlin intellectuals on the left could be ideologically committed to the Soviet agenda.

In June Günther Weisenborn left the *Ulenspiegel* and moved to Hamburg, where he became the director of the *Hamburger Kammerspiele* (1951–1953). In 1953 he published *Der Lautlose Aufstand: Bericht über die Widerstandsbewegung des deutschen Volkes 1933–1945* ("The Silent Uprising: A Report about the German Resistance Movement, 1933–1945"). Not surprisingly, Weisenborn's departure changed the journal. After he left the *Ulenspiegel* became an undisguised platform for Soviet propaganda. The journal's covers commented on neither the currency reform nor the Berlin Blockade.

In August ICD changed its name to Information Services Division (ISD), eliminating the term "control." At the same time, however, OMGUS tightened its control over the Western German press. That same month, the U.S. Feature Service—better known as Amerika Dienst—began to distribute news material developed by ISD to the German press. In theory, OMGUS exerted no pressure on the German editors to use the material, but the process had a covert dimension: ISD did not require that the German media attribute stories to Amerika Dienst, so it was impossible to know when the sources were in fact ISD dispatches.[44]

In September a new journal of caricature and satire appeared in the American sector. *Der Insulaner: Das Magazin für das Insel-Leben* ("The Islander: The Magazine for Insular Living") was edited by Günter Neumann, and its visual text had a distinctly pro-American, anti-Soviet, and anticommunist slant. The journal received rave reviews in the Western German press and was severely criticized in the East.[45] Neumann was one of the most successful satirists in Germany, and in December 1948 he launched the extraordinarily popular radio program *Günter Neumann und seine Insulaner* at RIAS. In 1949 Neumann opened the cabaret Der Insulaner, which became the most renowned nightclub in West Berlin, and wrote the script for one of the most biting and clever anticommunist propaganda films made under the Americans in Germany, *Nicht stören! Funktionärsversammlung* ("Do Not Disturb! Meeting in Progress").

In October 1948 Melvin J. Lasky launched *Der Monat* in the American sector of Berlin, which became, along with Sartre's *Les Temps Modernes,* one of the most important intellectual forums in postwar Europe. There has been much discussion in the past few years about the CIA's role in financing *Der Monat*.[46] OMGUS was clear about its political intentions of the journal: "This periodical is ... sponsored by Military Government in order to reorient the German people toward democratic values. *Der Monat* is designed to broaden the intellectual and cultural horizons of the Germans by providing serious discussion of important political, intellectual, and cultural issues by leading foreign and German writers and thinkers."[47] That same month, Sandberg returned his American license and took the *Ulenspiegel* to East Berlin, where it continued publication under a Soviet license.

Did the *Ulenspiegel* ever present an independent perspective on the incipient Cold War? The analysis of the visual text indicates that the journal criticized OMGUS and the American agenda for Germany and the world while keeping a quasi-hermetic silence on the Kremlin's foreign policy. This selective blindness was characteristic of the Communist press throughout the world, with which the *Ulenspiegel* shared tropes, symbols, themes, and clichés. In these publications, the Soviet Union enjoyed what François Furet refers to as "the ideological privilege of incarnating socialism."[48]

While the *Ulenspiegel* did not publish overt pro-Soviet propaganda, it reproduced many of the themes hammered on by the Comintern-dominated Communist press throughout Europe: the political agenda of Wall Street, the links between Wall Street, the military-industrial complex, and Washington, the relation between the United States and former Nazi industrialists, Pax Americana as a strategy for world domination, the imminent collapse of capitalism, heralded by the economic and social crises in the United States, American racism and violence against minorities, and the threat of a new world war if U.S. forces remained in Europe. By late 1947 the Soviet propaganda machine was calling for a "peace offensive" to counter the war psychosis allegedly being created by American political, business, and military leaders. The Marshall Plan was denounced as an imperialist maneuver to enslave Europe and revive an aggressive and expansionist Germany under American control. The revitalization of the Ruhr was interpreted as a ploy to strengthen German militarism, already growing due to what the Soviets estimated as the failure of denazification and demilitarization in the American zone.[49] All these arguments were reproduced in the *Ulenspiegel,* with the exception of the issue of American racism, on which the journal kept silent.

Many of the allegations made by the artists who collaborated in the *Ulenspiegel* were true. The American denazification project was indeed riddled with contradictions. Some American bankers and industrialists were interested in preserving the German bankers and industrialists of the Third Reich. Former Nazi industrialists, bankers, judges, bureaucrats, and military officers were in fact being reincorporated into Western Germany's new civil society en masse. As the United States increased its efforts to integrate Western Germany into Western Europe and to include it in the American-led anticommunist alliance, OMGUS modified its original denazification objectives. *Westintegration* and amnesty relied on a process of normalization, and reconstruction took priority over denazification, memory, and justice.[50] It was also true that the Truman Doctrine aimed to isolate the USSR, that Germany had particular importance in Cold War strategies, and finally, that postwar America was not the paradise of social peace and equality that American propaganda in Germany claimed.

Yet criticism of American policies was never matched by a similarly caustic analysis of Soviet policies. The *Ulenspiegel* did not discuss the failures of the denazification process in the East or the presence of former Nazi officers in politically sensitive positions in SMAD and the SED. That the journal's artists never referred to the extermination of two-thirds of the German Communists who had undergone exile in Moscow may have been due to lack of information. Yet their caricatures, so biting in their denunciation of the United States, never attacked Soviet foreign policy in Eastern Europe, simply ignoring the Communist coup in Czechoslovakia and its political aftermath. The *Ulenspiegel*'s artists ignored the Zhdanov campaign against modernism and the increasingly vociferous anti-Semitism (thinly disguised as "anticosmopolitanism") of SMAD and the SED. It would be difficult to deny that the *Ulenspiegel* took sides in the emerging Cold War.

The East Berlin Ulenspiegel

Once in East Berlin, the *Ulenspiegel* became an overt instrument of Communist propaganda. SMAD already had a satirical magazine, *Frischen Wind* ("Fresh Wind"), and the two journals coexisted for a time. The *Ulenspiegel*'s visual text stressed delinquency and terror in West Berlin and denounced the resurgence of militarism in Western Germany. In international matters, the *Ulenspiegel* attacked Charles De Gaulle, portraying him as an enemy of freedom, denounced the Marshall Plan, and demanded a new Germany and a new Europe independent from the United States. The caricatures in the East Berlin *Ulenspiegel* echoed themes in *Tägliche Rundschau*. Cartoons in the latter paper in late 1948 similarly denounced the alleged complicity between former Nazis and the United States, asserted that West German politicians were dogs controlled by their American masters, and implied that the West German elections had been rigged by General Clay.[51]

The *Ulenspiegel* repeatedly warned its readers about two imminent dangers—the catastrophic division of Germany, and the fracture of Europe into two irreconcilable camps. Sandberg's December 1948 collage "In Grunewald Forest" denounces Konrad Adenauer as an American agent intent on destroying Germany's fledgling democracy. The leader of the West German Christian Democrats, his shirt covered with Cold War headlines from the Western press, is about to chop down a slender tree representing the new German democracy. A woman strolling by is horrified and shouts, "Don't cut it, it is very young." Adenauer, referred to in the image as "the fury of war," replies, "The tree must go, I need fuel for conflict."[52] Looking beyond U.S.-German affairs, the journal urged the end—or at least the limitation—of America's influence over Europe. It denounced dollar diplomacy, attacked the Truman foreign policy and the Marshall Plan, critiqued the behavior of the Western powers in Greece, China, and Spain, and called for a European fraternity based on the exclusion of the United States.

In the visual text of the East Berlin *Ulenspiegel* the emergence of East Germany—the German Democratic Republic—is depicted as a glorious day. Elizabeth Shaw's cover for the December 1949 issue, "Land-Land! Deutsch-land!," published three months after the establishment of the GDR, shows its emergence as a symbol of national hope. Men, women, children, and animals aboard Noah's ark are lit by a rising sun, labeled *"Deutsche Demokratische Republik."* They gaze with hope at a dove of peace approaching their boat.[53]

The end of the East Berlin Ulenspiegel
Throughout 1949 the *Ulenspiegel*'s symbols and tropes did not change, but the attacks against the United States and West Germany became even more virulent. Wall Street capitalism and the American military and industry were allegedly bullying the world toward a new war in order to maximize their profits and gain global control. The American-German partnership in Western Germany, depicted as a happy embrace between a partner with dollars and a partner with swastikas, was said to inhibit the emergence of German democracy. In Sandberg's August 1949 "This is Democracy," General Clay is shown as the skipper of a warship. Germany is pictured as a tiny man with Adenauer's face at the helm of the ship's small life raft—labeled "Bonn." Clay says, "Naturally you are allowed to control our course!" In November, Sandberg drew a cover titled "Heuss: Take Courage and Jump into Independence." President Theodor Heuss, the first president of the Federal Republic of Germany, dressed in an Uncle Sam suit, is seen falling headfirst from atop a building labeled "Bonn" (a replica of the White House) toward the open jaws of an alligator wagging a tongue shaped like a dollar sign.[54]

Even as its anti-American rhetoric escalated, the *Ulenspiegel* did not comment on Soviet policy, either in Germany or abroad. The anti-Semitic and anticosmopolitan offensive that had begun in 1946 went unremarked, as did the persecution of non-Stalinist Communists, most of them Jews, in Hungary, Bulgaria, Czechoslovakia, and East Germany that began in September 1949, with the scandalous Rajk trial.[55]

Yet Sandberg tried to use the journal as a vehicle to reintroduce modern art into Eastern Germany. This agenda put him at odds with Soviet policy. In February 1949 he and Lieutenant Colonel Alexander Dymschitz, the head of the cultural division of SMAD's Office of Information, clashed in a public debate held at the Humboldt University in East Berlin, titled "Formalism and Realism in the Fine Arts." Sandberg defended modern art, the artistic tradition of European modernist art, and the work of Karl Hofer in particular. He stood up for the right of Communist artists to work for the revolution as they pleased, without formal restraints. Dymschitz accused Sandberg of being an agent of formalism, a reactionary, and a defender of a "rotten culture." Sandberg, he claimed, had ideological defects that led him to accept degeneration, decadence, and bourgeois values. The allegation was that Sandberg could not fit into the antifascist-democratic order that was to transform East Germany because he was not a true Marxist. In March 1949 Sandberg was labeled a "formal fascist" in an article appearing in *Neues Deutschland,* the SED's official mouthpiece.[56]

Sandberg complained about this criticism, recalling his ten years of martyrdom in Nazi concentration camps. He wanted to be recognized as a Marxist and as a realist artist, and argued that the SED's politicization of art would lead to the "triumph of mediocrity."[57] The official response came not from Dymschitz, who had been called back to Moscow, but from Stefan Heymann, a former Buchenwald inmate who had become an SED functionary. Heymann insisted that Sandberg was neither a Marxist nor a realist. Sandberg's public confrontation with Dymschitz had signaled the beginning of the end for the *Ulenspiegel.* Echoing OMGUS's tactic, SMAD cut the journal's paper supply. According to Sandberg, the pretext was that "in any people's democracy, there was room for only one satirical journal."[58]

The GDR authorities terminated Sandberg's licence in August 1950. According to Sandberg, the official reason was that the cover of the penultimate issue—two beer steins symbolizing the treaty of friendship between the German Democratic Republic and the Republic of Czechoslovakia—was not deemed to be serious enough. In Sandberg's opinion, however, the *Ulenspiegel* was not tolerated because he defended modern art and published modern literature in defiance of Soviet directives. When the *Ulenspiegel* closed, some of its staff went to work in *Frischen Wind,* while others, including Sandberg (following Ulbricht's advice), joined *Neues Deutschland.* In 1954 Sandberg became editor in chief of the monthly art journal *Bildende Kunst* ("Visual Arts"), the most important mouthpiece of Socialist Realism in East Germany.[59] He left *Bildende Kunst* in 1957, and began working on an autobiographical collection of etchings, which he finished in 1963. *Der Weg* ("The Path"), includes recollections of Sandberg's experiences in a Nazi concentration camp.

Beyond the *Ulenspiegel*

The period between the unconditional surrender of the Third Reich and the full emergence of the Cold War offered considerable political possibilities in the American zone and sector. Many American cultural officers welcomed German intellectuals from the left as ideal partners in the construction of an antifascist and democratic Germany. The initial American press policy was extraordinarily tolerant and allowed free debate regarding alternative German futures. Although in theory the American-licensed press was not allowed to criticize Allied occupation policies, in practice it did so. The result was a new German press that debated political issues and did not spare criticism of the American agenda for Germany.

The Soviets exploited this permissiveness in the immediate postwar years. They did not have to rely solely on complex intelligence operations to spread their worldview in West Berlin, because they could count on a unique asset within the American zone and sector—German left-wing intellectuals with impeccable anti-Nazi credentials who did not agree with the evolution of the American policy in Germany and who had been given a voice in the American-licensed press. The West Berlin *Ulenspiegel* was not a conventional Soviet intelligence asset, because Sandberg and Weisenborn never tried to pass for something that they were not. There is no evidence indicating that they were infiltrated into West Berlin to sabotage the American occupation and introduce an anti-American narrative, nor were they "bought" by the Soviets in any sense.

The Cold War did encroach on the possibility of independent political discourse. Stalin spoke of the menace of a new fascism—American imperialism—and Truman denounced the Soviet Union as a totalitarian state that continued the antidemocratic offensive of Nazism. This polarization reshaped American policy in Germany, as initial ideological openness gave way to distrust and later outright censorship and suppression of alternative viewpoints. The Soviet anti-American campaign was answered in kind, and OMGUS ensured that the media in West Germany became focused on anti-Soviet and anticommunsit propaganda. Neither the Soviets nor the Americans allowed real democratic freedom of expression in occupied Germany. On both sides, rhetoric about democracy and freedom of expression was qualified and subordinated to political needs.

Conclusion

The U.S. Army entered Germany with a vague administrative blueprint and a harshly contested political and economic agenda for the occupation of the country. Some in Washington prescribed a program of relentless and unforgiving denazification. Others were skeptical of denazification and demanded immediate political, economic, and industrial reconstruction. The initial policies of the American military government, therefore, often entailed complicated compromises and pragmatic solutions.

Yet in spite of these internal conflicts, the Americans succeeded in establishing a strong military government, exerting a monopoly of violence and information in the American zone and sector. The American political objective was the democratization of Germany, but OMGUS did not function as a democratic government. Rather, it established strict political limits in order to manage an agenda of radical regime change while battling the legacy of a proselytizing regime that had combined ideological fundamentalism with antidemocratic populism. At the same time, OMGUS had to avoid imposing alternatives so foreign that they would lead automatically to rejection and violence.

The American democratization agenda required a double perceptual revolution in West Germany: OMGUS had to change both how Germans perceived the Third Reich and how they perceived the United States. The emergence of the Cold War in 1947 introduced further challenges. The former German enemy became the new ally, and the Soviet ally the new archenemy. As a result, the intensity of the denazification process was lessened. By the end of the occupation, former Nazis were allowed to reenter German political and economic life, albeit within the ideological framework imposed first by OMGUS and later by the High Commissioner for Germany (HICOG).

American policy in occupied Germany did change during the 1945–1949 period. Yet certain aspects remained constant—the ban on Nazi propaganda remained in effect. The denazification of the visual sphere was arguably as vigorous in 1949 as in 1945.

Visual propaganda was an important element of the denazification and reeducation processes. The U.S. Army and OMGUS carried out a massive exercise of iconoclasm and censorship, and incorporated photography, film, caricature, and the fine arts into their psychological warfare program. Visual images served to introduce the German public in the American zone and sector to new social and ethical paradigms, and to spread new interpretations of Nazism and the war.

Atrocity propaganda, perhaps the most innovative American propaganda campaign in the immediate postwar period, was fundamentally a visual exercise. The most radical and successful denazification effort carried out by the Americans was the elimination of Nazi iconography and paraphernalia in the American zone and sector. This campaign of iconoclasm, coupled with the evidence of genocidal atrocities and the utter collapse of the Nazi government, challenged prevailing understandings of German national identity and drove home the catastrophic dimension of German military defeat.

OMGUS also used film to sell the American agenda for reconstruction to the population in the American zone and sector. Through fiction and documentary films, it introduced the notion of civil society as a form of nonviolent political interaction between equals, stressing the value of self-government, democratic participation, religious tolerance, freedom of the press, and community activism. No less important, OMGUS cultural officers encouraged the reemergence of modern art through private art associations, prizes, and cultural exchanges. The Americans allowed German modern artists to find their own styles and themes and helped them to establish bonds with American and Western European artists and intellectuals.

Simultaneously, though, the Cold War brought about a new kind of American censorship. Anti-Nazism ceased to be the exclusive litmus test for political acceptability. OMGUS began to purge Communists from its ranks and to exclude the West German procommunist left from the media in the American zone and sector. The case of the *Ulenspiegel* shows that OMGUS abandoned basic democratic principles and procedures in its attempt to impose a single, pro-American viewpoint in Western Germany. OMGUS was trapped in an impossible dilemma. It could not counter the *Ulenspiegel* with a parallel publication that was overtly controlled by the Americans, because this would delegitimize its claim of press freedom. At the same time, it could not tolerate the ongoing diffusion of a visual text that was increasingly anti-American and pro-Soviet, especially because many of the criticisms voiced by the *Ulenspiegel* concerning the failure of denazification in the American zone and sector had an element of truth. Eventually, OMGUS cut the *Ulenspiegel*'s supply of paper, prompting its departure to East Berlin, and issued a license to Günter Neumann, a satirist who had collaborated actively with the Americans since the beginning of the occupation, to publish a new satirical journal, *Der Insulaner*. By late 1948 political caricature in West Berlin was monolithically pro-American and anti-Soviet.

The Americans did not arrive to Germany with a clear cultural agenda. Their policy was developed on the ground, in the context of the confrontation with the USSR, the competition with the rest of the Western Allies, the need to satisfy the cultural aspirations of the German population, and ongoing cultural wars in the United States and within OMGUS. Many of the most innovative cultural initiatives were carried out by individuals or small groups, often working on the fringes of the military government and maintaining a low profile in order to avoid confrontation with the War Department and the U.S. Congress, both of which were dominated by radical conservatives.

American domestic politics played an important role in defining the tropes of American anticommunist propaganda. During the period 1945–1949, the United States was divided along racial lines, with significant sectors of the political and military establishment promoting an antiegalitarian and racist agenda. President Truman's domestic and foreign policies were constrained by a socially conservative Congress, and the War Department dragged its feet on the desegregation of the American military. Soviet anti-American propaganda exploited these contradictions, pointing to the disparity between the purported American belief in democracy, equality, and civil rights and the realities of institutionalized racism.

The campaign against *The Brotherhood of Man* waged by Congress, the U.S. armed forces, and the War Department illustrates how racism shaped the American response to Soviet anti-American propaganda. Yet despite the persistence of institutionalized racism in the United States, the American political establishment demanded that Germans radically change their perception of the Nazi past. OMGUS tried to alter the way Germans remembered Nazism by introducing visual evidence of Nazi criminality and suppressing pro-Nazi propaganda. The American confrontation policy and atrocity propaganda campaign were attempts to preclude nostalgia for the Nazi experience through visual means, irrevocably linking Nazism to atrocity and horror.

In 1945 the Americans faced the political dilemma of deciding who to hold responsible for the Nazi atrocities—Hitler and the upper echelons of the Nazi leadership or the entire German population. The confrontation policy failed to create collective guilt, and it ran the risk of fomenting pro-Soviet sympathies. With the emergence of the Cold War, the campaign was abandoned. The new goal was to win the hearts and minds of the German population and to establish an American–West German alliance. The Nazi leadership was put on trial, but the day-to-day complicity of the German population in their crimes was no longer emphasized. OMGUS did not create a culture of remembrance of the Nazi genocide based on the concept of German collective guilt. This was a political decision taken by the Americans, and not simply a German psychological reaction to trauma.

Nonetheless, the confrontation policy succeeded in creating a new image of the Third Reich grounded on horror, murder, and torture, and this helped to delegitimize the Nazi regime. After 1945, Germans could no longer identify themselves publicly with the Nazi conception of race, nation, and national *Kultur*.

OMGUS also succeeded in attracting West Germans to a democratic model of society. The first free elections in West Germany took place on August 14, 1949, less than five years after the unconditional surrender of the Third Reich. Although the political philosophies of the two leading candidates, the Social Democrat Kurt Schumacher and the Christian Democrat Konrad Adenauer, differed greatly, both men were fully committed to West Germany's integration with Western Europe and to maintaining a strong alliance with the United States. Fewer than 6 percent of the votes cast went to the Communist Party. The United States had been able to win over the West German population to the political and economic tenets of capitalist democracy.

This was in large part a victory for American propaganda. Psychological warfare is the main weapon in the battle for perceptions, and OMGUS used it masterfully. The American military government was able to persuade the West Germans that the American model of society was best and that the consequences of rejecting it would be dire. The Berlin Blockade was a massive Soviet blunder, providing the Americans an opportunity to showcase both their technological and economic power and their willingness and capacity to help the population of West Berlin. The American-British airlift was itself a massive operation of visual propaganda, though that was not its primary aim. The images of airplanes delivering coal, food, and medicine to the West Berliners became symbols of the Western commitment to a noncommunist Germany. The American anticommunist propaganda offensive succeeded in winning the allegiance of the West German population even before there was massive tangible evidence of the success of the liberal capitalist model in Germany. From its inception, then, the Federal Republic of Germany was anticommunist and committed to democracy, European integration, and collaboration with the United States.

Yet changes in West German political culture were gradual. In 1951 American intelligence polls determined that 44 percent of West Germans still considered the Third Reich to have been the best period in modern German history, while only 2 percent deemed the postwar years the best. Defeat and occupation did not abolish anti-Semitism either—in 1952, 37 percent of West Germans stated that it was better not to have Jews in Germany. Anecdotal evidence—private conversations, jokes, and graffiti—confirm that OMGUS was not able to eradicate completely the subculture of Nazi nostalgia.[1] Four years of military occupation were not enough to invalidate basic assumptions inculcated by twelve years of intense Nazi indoctrination.

The resilience of racist and antidemocratic sentiments during the first decade of the Federal Republic of Germany, however, does not indicate that the American denazification and democratization effort was a failure. In 1945 the U.S. Army brought about radical regime change in Germany. The change was abrupt and imposed from above. Yet new rules could not instantly transform a country emerging from a brutal, racist, and murderous dictatorship into a civil society based on respect for human rights, tolerance, and political freedom. Societal change was gradual. The transformation of West Germany was an endogenous process that took place within German society at its own tempo. The influence of OMGUS was profound, but the American military government did not "Americanize" Germany. The Federal Republic of Germany developed a parliamentary system, a strong welfare state, and a political and intellectual culture that did not replicate the American model.

After 1949 questions of morality, guilt, and responsibility for the Nazi genocide languished. Political success and the legitimation of West Germany as a democratic and normalized state were, however, apparently incompatible with the exploration of the past. The candidates in the 1949 election avoided a frank discussion of the horrors of Nazism.[2] West German politicians instead focused on establishing solid economic and political relations with Western Europe and recuperating the pre-Nazi bourgeois traditions. For seventeen years the country was led by the conservative Christian Democratic Union, first under Chancellor Konrad Adenauer (1949–1963) and then under Chancellor Ludwig Erhard (1963–1966). During this period, the country progressed economically within a rigid social structure that tended to stifle discussion of the Nazi past. In 1966 conservatives and Social Democrats formed the "Grand Coalition," but this political experiment was interrupted by the events of 1968, as West German students joined an international revolt, voicing their dissatisfaction with the West German state and society. The student revolt shook West Germany and spawned dramatic political, social, and cultural changes. The younger generation, born after Nazism, demanded the end of the era of amnesia and the public discussion of individual responsibility for Nazi crimes. A collective discussion of the Nazi experience finally began in earnest in West Germany. In 1969 Willy Brandt became the first Social Democratic chancellor since 1928, and he inaugurated an era of social-liberal reform. In the 1970s West Germans developed the concept of "constitutional patriotism," which offered an alternative to the political tenets of the Adenauer era. Constitutional patriotism attracted Germans preoccupied with civil liberties and the expansion of democracy in Germany. The Social Democratic/Liberal coalition remained in power until 1982.

While many West German intellectuals and writers had attempted to come to terms with the Nazi genocide and the question of German guilt, it was an American soap opera that finally changed the popular perception of the Holocaust in West Germany. The power of images in politics is exemplified by the effect of *Holocaust,* an NBC television miniseries screened in West Germany in 1979. This dramatization of the Nazi genocide triggered an unprecedented debate about the Nazi past. For the first time, West Germans were able to relate specifically to the Jewish victims of Nazism. So intense was the public reaction that the Bundestag decided to abrogate the statute of limitations on war crimes.[3] The Holocaust became integral to West German self-understanding. The new public debate about the Third Reich achieved what the American atrocity propaganda policy had attempted in 1945—to confront the German population with the extent of Nazi criminality. In fact, it is difficult to imagine the politics of memory in postwar West Germany without the American visual propaganda policies in 1945.

The military occupation of Germany also had an impact on the United States. President Roosevelt had taken great care to dissociate American intervention in World War II from the Jewish question. American wartime propaganda had ignored the horrors of the Nazi concentration and extermination camps. After the war, however, the liberation of those camps was used to construct a new justification for both American involvement in the war and the prolonged occupation of Germany. The defeat of the Third Reich and the successful moral and political reformation of West Germany acquired a unique symbolic value in the American national narrative. World War II became the paradigmatic "good war."

Moreover, the German experience helped the American intelligence community understand the nature of cultural warfare and established the model for later clandestine cultural interventions by the Central Intelligence Agency. Many of the ICD cultural warriors, such as Melvin Lasky and Michael Josselson, became important players in those CIA campaigns. CIA operatives learned to use private foundations, to infiltrate intellectual circles, to organize cultural events, and to channel funds, in order to foster the American agenda of the Cold War.

It would be simplistic to draw a causal relation between the OMGUS period and contemporary Germany, a unified country in a post–Cold War world. Yet it would be equally problematic to underestimate the importance of the occupation in constructing the parameters of the possible, the permissible, and the desirable in West Germany. American occupation forces had the goal of eliminating militarism and extreme nationalism in the American zone and sector, and to a great extent they succeeded. The four years of military occupation did not transform West Germans into ardent pacifists, but they created the political context that allowed the emergence of a pacifist nation. It is only now, more than sixty years after World War II, that the German government is debating whether Germany should have a stronger military presence abroad. In the years following the surrender of the Third Reich, West Germans developed a very particular kind of nonnationalist national identity. Many were uncomfortable with explicit signs and icons of nationalism, preferring to embrace a supranational identification as Europeans.[4] And reunification did not dispel the ghosts of the past. In Germany the Holocaust continues to shape the choices and attitudes of the present.

Notes

Introduction

1. For a discussion of American psychological warfare, see Alfred H. Paddock Jr., *U.S. Army Special Warfare: Its Origins: Psychological and Unconventional Warfare, 1941–1952* (Washington, DC: National Defense University Press, 1982).

2. Harold Zink, *The United States in Germany, 1944–1955* (Princeton: D. Van Nostrand, 1957) and *American Military Government in Germany* (New York: MacMillan, 1947); John Gimbel, *The American Occupation of Germany* (Stanford: Stanford University Press, 1968) and *A German Community under American Occupation: Marburg 1945–1952* (Stanford: Stanford University Press, 1961).

3. Zink, *United States in Germany*, 234.

4. This trend is evident in Karin von Hippel, *Democracy by Force: U.S. Military Intervention in the Post–Cold War World* (Cambridge: Cambridge University Press, 2000); Tony Smith, *America's Mission: The United States and the Worldwide Struggle for Democracy in the Twentieth Century* (Princeton: Princeton University Press, 1994); and Andrew J. Bacevich, *American Empire: The Realities and Consequence of U.S. Diplomacy* (Cambridge: Harvard University Press, 2002).

5. Some notable examples are Dagmar Barnouw, *Germany 1945: Views of War and Violence* (Bloomington: Indiana University Press, 1996); Clare Flanagan, *A Study of German Political-Cultural Periodicals from the Years of Allied Occupation, 1945–1949* (Lewiston: Edwin Mellen Press, 2000); Jessica C. E. Gienow-Hecht, *Transmission Impossible: American Journalism as Cultural Diplomacy* (Baton Rouge: Louisiana State University Press, 1999); Brigitte J. Hahn, *Umerziehung durch Dokumentarfilm? Ein Instrument amerikanischer Kulturpolitik im Nachkriegsdeutschland (1945–1953)* (Munster: Lit Verlag, 1997); Larry Hartenian, *Controlling Information in U.S. Occupied Germany, 1945–1949: Media Manipulation and Propaganda* (Lewiston: Edwin Mellen Press, 2003); Johannes Hauser, *Neuaufbau der westdeutschen Filmwirtschaft 1945–1955* (Freiburg: Centaurus-Verlagsgesellschaft, 1989);Yule F. Heibel, *Reconstructing the Subject: Modernist Painting in Western Germany, 1945–1950* (Princeton: Princeton University Press, 1995); Uta G. Poiger, *Jazz, Rock, and Rebels: Cold War Politics and American Culture in a Divided Germany* (Berkeley: University of California Press, 2000); Hanna Schissler, ed., *Germany: The Miracle Years* (Princeton: Princeton University Press, 2001); Barbie Zelizer, *Remembering To Forget: Holocaust Memory through the Camera's Eye* (Chicago: University

of Chicago Press, 1998); Cora Sol Goldstein, "Power and the Visual Domain: Images, Iconoclasm, and Indoctrination in American Occupied Germany, 1945–1949," PhD diss., University of Chicago, 2002; and Michael Hoenisch, Klaus Kämpfe, and Karl-Heinz Pütz, *USA und Deutschland: Amerikanische Kulturpolitik 1942–1949* (Berlin: J. F. Kennedy–Institut für Nordamerikastudien, 1980).

6. See Volker R. Berghahn, *America and the Intellectual Cold Wars in Europe* (Princeton: Princeton University Press, 2001); David Caute, *The Dancer Defects: The Struggle for Cultural Supremacy during the Cold War* (Oxford: Oxford University Press, 2005); Victoria de Grazia, *Irresistible Empire: America's Advance Through 20th-Century Europe* (Cambridge: Harvard University Press, 2005); Giles Scott-Smith and Hans Krabbendam, eds., *The Cultural Cold War in Western Europe* (London: Frank Cass, 2003); Ralph Willett, *The Americanization of Germany, 1945–1949* (London: Routledge, 1989); Richard Pells, *Not Like Us: How Europeans Have Loved, Hated, and Transformed American Culture since World War II* (New York: Basic Books, 1997); and Frances Stonor Saunders, *The Cultural Cold War: The CIA and the World of Arts and Letters* (New York: New Press, 2000).

7. For further discussion on the importance of visual perception in shaping behavior, see W. J. T. Mitchell, ed. *Art in the Public Sphere* (Chicago: University of Chicago Press, 1992); David Freedberg, *The Power of Images: Studies in the History and Theory of Response* (Chicago: University of Chicago Press, 1989); John Berger, *About Looking* (New York: Vintage, 1980); Susan Sontag, *Regarding the Pain of Others* (New York: Picador, 2003); and Susan Sontag, *On Photography* (New York: Picador, 2001).

8. See Thomas Doherty, *Projections of War: Hollywood, American Culture, and World War II* (New York: Columbia University Press, 1993), and Clayton R. Koppes and Gregory D. Black, *Hollywood Goes to War: How Politics, Profits, and Propaganda Shaped World War II Movies* (New York: Basic Books, 1987).

9. See Joseph W. Bendersky, "From Cowards and Subversives to Aggressors and Questionable Allies: U.S. Army Perceptions of Zionism since World War I," *Journal of Israeli History* 25, no. 1 (2006): 107–29, and "The Absent Presence: Enduring Images of Jews in United States Military History," *American Jewish History* 89, no. 4 (2001): 411–36.

10. Joseph W. Bendersky, *The "Jewish Threat": Anti-Semitic Politics of the U.S. Army* (New York: Basic Books, 2000), 14.

11. Ibid., 239, 197–287.

12. Paddock, *U.S. Army Special Warfare*, 9.

13. Ibid., 32.

14. For further discussion of the OSS, see Charles A. Thomson, *Overseas Information Service of the United States Government* (Washington, DC: Brookings Institution, 1948); Paul M. A. Linebarger, *Psychological Warfare* (Washington, DC: Combat Forces Press, 1948); Bradley F. Smith, *The Shadow Warriors: O.S.S. and the Origins of the C.I.A.* (New York: Basic Books, 1983); Barry M. Katz, *Foreign Intelligence: Research and Analysis in the Office of Strategic Services, 1942–1945* (Cam-

bridge: Harvard University Press, 1989); Michael Warner, *The Office of Strategic Services: America's First Intelligence Agency* (Washington, DC: CIA Public Affairs Office, 2000); and Thomas W. Braden and Stewart Alsop, *OSS—Sub Rosa: The OSS and American Espionage* (New York: Reynal and Hitchcock, 1946).

15. During World War II, the General Staff Divisions of the U.S. Army were G-1, G-2, G-3, G-4, G-5, and G-6. G-1 dealt with personnel matters—human resources (military and civilian), manning, discipline, and personal services. G-2 was responsible for intelligence and security; G-2 personnel advised staff officers on intelligence and counterintelligence matters in their areas of operation. G-3 was in charge of organization and training, and included combat development, combat health support, and the elaboration of tactical doctrine. G-4, the logistics division, coordinated supply, transport, maintenance, and quartering. G-5 was the civil affairs division, concerned with civil-military operations and civil-military cooperation. G-6 was responsible for all matters concerning communications. See Mark Skinner Watson, *United States Army in World War II* (Washington, DC: Center of Military History United States Army, 1991).

16. In 1943 the United States and the UK formed the Chief of Staff to the Supreme Allied Commander (COSSAC) to plan the invasion of continental Europe and the occupation of Germany. The occupation plans were limited, aimed at securing and holding strategic areas in Germany after the surrender of the Wehrmacht. In January 1944 COSSAC became part of SHAEF. The German Section of the Civil Affairs Staff of COSSAC became the G-5 Division of SHAEF and designed the structure of the future military government in Germany. In March 1944 the German Country Unit in SHAEF wrote the first *Handbook for Military Government in Germany,* and that June the Posthostilities Planning Section of SHAEF prepared Operation Eclipse. Eclipse outlined the tasks to be carried out by the military government, including provisions for surrender, sanctions, disarmament, disposal of enemy material, war criminals, displaced persons and prisoners of war, and intelligence. Eclipse also covered the control of paramilitary forces, of transportation and communications, of the police, of administrative and political structures, and of public information and media. See Kenneth O. McCreedy, "Planning the Peace: Operation Eclipse and the Occupation of Germany," *Journal of Military History* 65, no. 3 (July 2001): 713–39.

17. After V-E Day, the united American-British command dissolved. To comply with the Yalta agreement, the British and American elements in SHAEF were placed under independent national commands. This required a change in the structure of the U.S. Army in Europe. Operationally, American forces were under SHAEF command—staffed by American and British officers—and administratively they were subordinated to the headquarters of the European Theater of Operation, U.S. Army (ETOUSA). Eisenhower was the commanding general of both SHAEF and ETOUSA. On July 1, 1945, ETOUSA was renamed U.S. Forces, European Theater (USFET); its headquarters remained in Frankfurt. SHAEF was dissolved on July 14, 1945. In November 1945 Eisenhower returned to Washington, D.C., as chief

of staff of the U.S. Army. General George S. Patton assumed command of USFET but died before the end of the year; he was replaced by General Joseph T. McNarney. USFET was replaced in March 1947 by the European Command (EUCOM). OMGUS ceased to exist on December 5, 1949. For the detailed discussion of the early American occupation plans and the organization of the military government in Germany, see Oliver J. Frederiksen, *The American Military Occupation of Germany, 1945–1953* (United States Army, Europe, Headquarters, Historical Division, 1953), 1–42.

18. Ibid., 14, 31–34.

19. Daniel Lerner, "Afterword—The Psychological Warfare Campaign against Germany: D-Day to V-E Day," in Victor Margolin, ed., *Propaganda: The Art of Persuasion: World War II* (New York: Chelsea House, 1975), 3.

20. Earl Zimke, *The U.S. Army Occupation of Germany, 1944–1946* (Washington, DC: Center of Military History, U.S. Army, 1975), 45–46.

21. All major German cities were from 50 to 75 percent destroyed. Berlin, the largest, was 75 percent destroyed, and Frankfurt, where the Supreme Headquarters of the U.S. Army in Germany were located, was 60 percent destroyed. Frederiksen, *American Military Occupation of Germany*, 12.

22. On August 5, 1944, the Joint Chiefs of Staff instructed SHAEF to create the U.S. Group Control Council, Germany (USGCC), which became the American element of the Allied Control Council. USGCC was commanded by General Eisenhower. The Allied Control Council functioned from May 8 to October 1, 1945. On October 1, 1945, USGCC became OMGUS. See Frederiksen, *American Military Occupation of Germany*, 1–42.

23. See Rebecca Boehling, *A Question of Priorities: Democratic Reforms and Economic Recovery in Postwar Germany* (New York: Berghahn Books, 1996), 15–40, and Earl F. Ziemke, "Improvising Stability and Change in Postwar Germany," in Robert Wolfe, ed., *Americans as Proconsuls: United States Military Government in Germany and Japan, 1944–1952* (Carbondale: Southern Illinois University Press, 1984), 52.

24. Carl J. Friedrich, "Military Government and Dictatorship," *Annals of the American Academy of Political and Social Science* 267 (January 1950): 1. See also Friedrich, "Military Government as a Step toward Self-Rule," *Public Opinion Quarterly* 7, no. 4 (Winter 1943): 527–41. Friedrich later helped draft the constitution of the Federal Republic of Germany. For biographical details, see Papers of Carl J. Friedrich, Harvard University Library.

25. The text of JCS 1067 can be found in U.S. Department of State, *Germany 1945–1949: The Story in Documents* (Washington, DC: U.S. Government Printing Office, 1950), 21–33.

26. Lucius D. Clay, *Decision in Germany* (Melbourne: William Heinemann, 1950), 87; Gimbel, *American Occupation of Germany*, 51. For analysis of German grassroots democracy and self-government, see Boehling, *Question of Priorities*, 156–210, and Konrad H. Jarausch, *After Hitler: Recivilizing Germans, 1945–1955* (Oxford: Oxford University Press, 2006), 132–35.

27. Edward C. Breitenkamp, *The U.S. Information Control Division and Its Effect on German Publishers and Writers 1945 to 1949* (Grand Forks, ND: University Station, 1953), 1.

28. See David Pike, *The Politics of Culture in Soviet-Occupied Germany, 1945–1949* (Stanford: Stanford University Press, 1992); Bernard Genton, *Les alliés et la culture: Berlin, 1945–1946* (Paris: Presses Universitaires de France, 1988); Norman M. Naimark, *The Russians in Germany: A History of the Soviet Zone of Occupation, 1945–1949* (Cambridge: Harvard University Press, 1997); Reinhard Müller, *Menschenfalle Moskau: Exil und Stalinistische Verfolgung* (Hamburg: Hamburger Edition, 2001); and Jeffrey Herf, *Divided Memory: The Nazi Past in the Two Germanys* (Cambridge: Harvard University Press, 1997).

29. Brewster S. Chamberlain, *Kultur auf Trümmern: Berliner Berichte der amerikanischen Information Control Section Juli–Dezember 1945* (Stuttgart: Deutsche Verlags-Anstalt, 1979), 16.

30. The Soviet strategy in Germany followed a model developed by Willi Münzenberg (1889–1940), a leading propagandist for the German Communist Party during the Weimar era, which made simultaneous use of overt and covert propaganda programs. Münzenberg organized a vast network of ostensibly nonpartisan associations—committees, literary journals, bookstores, cinema clubs, study groups, "independent theaters," and debate associations—that would spread the Communist message without apparent links to the Communist Party. These front organizations (or "Innocent Clubs," as Münzenberg called them) sought to attract, persuade, and coopt liberal and leftist intellectuals ("innocents") all over the world. Steven Koch, *Double Lives: Stalin, Willi Münzenberg and the Seduction of the Intellectuals* (New York: Enigma Books, 2004), 15–17. For an account of the impact of Münzenberg's doctrine of strategic propaganda, see François Furet, *Le passe d'une illusion: Essai sur l'idee communist au XX siècle* (Paris: Robert Laffont, 1995).

31. Pike, *Politics of Culture,* 80–135.

32. The text of JCS 1779 can be found in U.S. Department of State, *Germany 1945–1949,* 33–42.

33. Saunders, *Cultural Cold War,* 7–32. Also see Giles Scott-Smith, "A Radical Democratic Political Offensive: Melvin J. Lasky, Der Monat, and the Congress for Cultural Freedom," *Journal of Contemporary History* 35, no. 2 (April 2000): 263–80.

34. See Hartenian, *Controlling Information;* Gienow-Hecht, *Transmission Impossible;* and Flanagan, *Study of German Political-Cultural Periodicals.*

35. See Alan Bance, ed., *The Cultural Legacy of the British Occupation in Germany* (Stuttgart: Hans-Dieter Heinz, 1997), and Frances A. Rosenfeld, "The Anglo-German Encounter in Occupied Hamburg, 1945–1950," PhD diss, Columbia University, 2006.

36. Genton, *Les alliés et la culture,* 279–312.

Chapter One

1. American intelligence records referred to the practice as "atrocity propaganda" or "atrocity information policy." *Atrocities: A Study of German Reactions,* Supreme Headquarters Allied Expeditionary Force, Psychological Warfare Division, Intelligence Section, June 21, 1945, 7. Robert H. Abzug's pioneering study offers a historical account of the American liberation of Nazi concentration camps, and describes their impact on the liberators and on the American public. Robert H. Abzug, *Inside the Vicious Heart: Americans and the Liberation of Nazi Concentration Camps* (Oxford: Oxford University Press, 1985).

2. Sybil Milton, "Confronting Atrocities," in *Liberation 1945* (Washington, DC: United States Holocaust Museum, 1995), 61.

3. Harold Marcuse, "Nazi Crimes and Identity in West Germany: Collective Memories of the Dachau Concentration Camp, 1945–1990," PhD diss., University of Michigan, 1992, 83.

4. United States Holocaust Memorial Museum (USHMM) Photo Archives. For more on the use of atrocity photographs, see Cornelia Brink, *Ikonen der Vernichtung: Öffentlicher Gebrauch von Fotografien aus nationalsozialistischen Konzentrationslagern nach 1945* (Berlin: Akademie Verlag, 1998), and Dagmar Barnouw, *Germany 1945: Views of War and Violence* (Bloomington: Indiana University Press, 1996).

5. David Culbert, "American Film Policy in the Re-education of Germany after 1945," in Nicholas Pronay and Keith Wilson, eds., *The Political Re-Education of Germany and Her Allies after World War II* (Totowa, NJ: Barnes and Noble Books, 1985), 177.

6. Quoted in Oliver J. Frederiksen, *The American Occupation of Germany 1945–1953* (United States Army, Europe, Headquarters, Historical Division, 1953), 9; Eugene Davidson, *The Death and Life of Germany* (New York: Alfred A. Knopf, 1959), 54.

7. The text of JCS 1067 can be found in U.S. Department of State, *Germany 1945–1949: The Story in Documents* (Washington, DC: U.S. Government Printing Office, 1950), 21–33.

8. The U.S. Army liberated the following camps in Germany: Ohrdruf (Buchenwald subcamp) on April 4, 1945 (89th Infantry, 4th Armored); Ahlem (Neuengamme subcamp), April 10 (84th Infantry); Langenstein (Buchenwald subcamp), April 11 (83rd Infantry); Dora-Mittelbau, April 11 (3rd Armored, 104th Infantry); Buchenwald, April 11–12 (6th Armored, 80th Infantry);Halberstadt-Zwieberge (Buchenwald subcamp), April 12–17 (8th Armored); Salzwedel (Neuengamme subcamp),April 14 (84th Infantry); Leipzig-Schoenfeld (Buchenwald subcamp), April 14 (2nd Infantry); Leipzig-Thelda (Buchenwald subcamp), April 19 (69th Infantry); Flossenburg subcamp, April 20–21 (65th Infantry); Flossenburg, April 23 (90th Infantry); Landsberg (Dachau subcamp), April 27–28 (103rd Infantry, 12th Armored, 10th Armored, 101st Airborne); Dachau subcamps, April 28–29 (4th Infantry); Dachau, April 29 (20th Armored, 42nd Infantry, 45th Infantry); Kaufering camps (Dachau sub-

camps), April 29–30 (63rd Infantry, 36th Infantry); Dachau subcamps, May 2–3 (14th Armored); Wöbbelin (Neuengamme subcamp), May 3 (8th Infantry, 82nd Airborne); Dachau subcamp, May 3–4 (99th Infantry); Ebensee (Mauthausen subcamp), May 4–5 (80th Infantry); Gusen (Mauthausen subcamp), May 5 (11th Armored); Gunskirchen (Mauthausen subcamp), May 5–6 (71st Infantry); Mauthausen, May 6, 1945 (11th Armored); and Falkenau an der Eger (Flossenblurg subcamp), May 7 (9th Armored, 1st Infantry).

9. Linton stressed the incomparability of the camp's brutality to anything he had encountered in combat in an interview with the author, New York, March 22, 2000. Leonard Linton, "Kilroy Was Here" (unpublished manuscript, 1997), 127. See also "Crime and Punishment," *Army Talks* 4, no.9, July 10, 1945, 2, and the United States Holocaust Memorial Museum Archives (USHMM): RG 09. A similar sense comes from a pamphlet written by the PWB Section, 7th Army, *Dachau*, 1945, and distributed widely to American soldiers.

10. See Abzug, *Inside the Vicious Heart;* Charles Whiting, *Patton's Last Battle* (New York: Stein and Day, 1987); and Martin Blumenson, *The Patton Papers, 1940–1945* (Boston: Houghton Mifflin, 1974).

11. "Short Memories and Nice People," *Yank: The Army Weekly,* October 26, 1945.

12. "This Is Why We Fight," reprinted in Robert Abzug, ed., *GIs Remember: Liberating the Concentration Camps* (Washington, DC: National Museum of American Jewish Military History, 1994), 38.

13. "Crime and Punishment," 4. For an analysis of the impact of the Holocaust on the American press and the American government, see Deborah Lipstadt, *Beyond Belief: The American Press and the Coming of the Holocaust, 1933–1945* (New York: Free Press, 1993); David S. Wyman, *The Abandonment of the Jews: America and the Holocaust, 1941–1945* (New York: Pantheon, 1984); and David S. Wyman, *America and the Holocaust* (New York: Garland Publishing, 1990).

14. Gallup Poll #346, May 5, 1945.

15. Milton, "Confronting Atrocities," 55; "Crime and Punishment," 2.

16. Dwight D. Eisenhower, telegram to George C. Marshall, April 19, 1945, reprinted in *Atrocities and Other Conditions in Concentration Camps in Germany,* Report of the Committee Requested by General Dwight D. Eisenhower through the Chief of Staff, General George C. Marshall, to the Congress of the United States, Presented by Mr. Barkley, May 15, 1945 (Washington, DC: U.S. Government Printing Office), 1. It is remarkable that Eisenhower referred to those held in the concentration camps as "political prisoners" and did not mention the racist element of the Nazi genocide. However, Jews were a minority—about one-fifth—of the inmates liberated by American troops in Germany. See Peter Novick, *The Holocaust in American Life* (Boston: Houghton Mifflin, 1999), 65.

17. The group was led by Senator Alben Barkley of Kentucky, the Senate majority leader and later Truman's vice president and also included Senators Leverett Saltonstall (Massachusetts), Walter George (Georgia; acting chairman of the Com-

mittee on Foreign Relations), Kenneth Wherry (Nebraska), Elbert Thomas (Utah; chairman of the Committee on Military Affairs), and C. Wayland Brooks (Illinois) and Representatives John Vorys (Ohio), Ewing Thomason (Texas; senior member of the Military Affairs Committee), James Richards (South Carolina), Edouard Izac (California), James Mott (Oregon), and Dewey Short (Missouri).

18. The journalists were Joseph Pulitzer, *St. Louis Post-Dispatch;* Julius Ochs Adler, *New York Times;* L. K. Nicholson, *New Orleans Time-Picayune;* Stanley High, *Reader's Digest;* William Nichols, *This Week;* Norman Chandler, *Los Angeles Times;* Amon Carter, *Fort Worth Star-Telegram;* John Randolph Hearst, Hearst Publications; Beverly W. Smith Jr., *American Magazine;* Gideon Seymour, *Minneapolis Star-Journal;* Ben Hibbs, *Saturday Evening Post;* Benjamin M. McKelway, *Washington Post;* Walker Stone, Scripps Howard Newspaper Alliance; M. E. Walter, *Houston Chronicle;* Malcolm Bingay, *Detroit Free Press;* Duke Schoop, *Kansas City Star;* E. C. Dimitman, *Chicago Sun;* and William Chennery, *Collier's Magazine.*

19. Pulitzer, quoted in Abzug, *Inside the Vicious Heart,* 132.

20. "The Pictures Don't Lie," *Stars and Stripes,* April 26, 1945, 2; "Crime and Punishment," 2; "Indisputable Proof," *News Chronicle,* April 19, 1945, 1; "Torture Camps: This Is the Evidence," *Daily Mail,* April 19, 1945, 2. See also Barbie Zelizer, *Remembering to Forget: Holocaust Memory through the Camera's Eye* (Chicago: University of Chicago Press, 1998), 92; Milton, "Confronting Atrocities"; PWB Section, Seventh Army, *Dachau, Concentration Camp and Town,* 1 (Dachau Museum and Memorial Archive); "Report on German Murder Mills," *Army Talks* 4, no. 9 (July 10, 1945): 7; "Congress Irate at Atrocity Reels," *Variety,* May 2, 1945, 2.

21. "Mr. Pulitzer's First Article from Paris," *St. Louis Post-Dispatch,* April 29, 1945; "War Department Releases Atrocity Films for Nation to See," *St. Louis Post-Dispatch,* June 7, 1945; Joseph Pulitzer, *A Report to the American People* (St. Louis: St. Louis Post-Dispatch, 1945), 102.

22. Luther Evans, Librarian of Congress, letter to Raymond Brandt, *St. Louis Post-Dispatch,* January 25, 1946 (Library of Congress Archives: German Atrocity Pictures, Central File System, #786); Daniel W. Pfaff, *Joseph Pulitzer and the Post-Dispatch* (Pennsylvania: Pennsylvania State University Press, 1991).

23. Jennifer L. Mnookin, "The Image of Truth: Photographic Evidence and the Power of Analogy," *Yale Journal of Law and the Humanities* 10, no. 1 (Winter 1998): 66. Photographic evidence may be falsified in various ways. The context in which a photograph was made can be misrepresented, a caption can twist its meaning, or the image itself can be tampered with. The crude techniques by which figures were eliminated from official Soviet photographs have evolved, replaced by more sophisticated retouching and digital imaging. The most common distortion of context is staging, aimed at emphasizing attention-getting and emotion-triggering images. See Susan Sontag, *Regarding the Pain of Others* (New York: Picador, 2003) (quotation, p. 26); Richard Dorment, "On Hanging," *New York Review of Books,* September 22, 2005.

24. Kevin Grant, *A Civilised Savagery: Britain and the New Slaveries in Africa, 1884–1926,* London: Routledge, 2005), 42; Sontag, *Regarding the Pain of Others,* 83.

25. See Paul Messaris, *Visual Persuasion: The Role of Images in Advertising* (Thousand Oaks: Sage Publications, 1997), 137; Kathleen Collins, "Portraits of Slave Children,"*History of Photography: An International Quarterly* 9, no. 3 (1985): 187–210; and William A. Frassanito, *Gettysburg: A Journey in Time* (New York: Charles Scribner's Sons, 1975).

26. See Frances Fralin, *The Indelible Image: Photographs of War—1846 to the Present* (New York: Harry N. Abrams, 1985); Jorge Lewinski, *The Camera at War: A History of War Photography from 1848 to the Present Day* (New York: Simon & Schuster, 1980); and George H. Roeder, *The Censored War: American Visual Experience during World War Two* (New Haven: Yale University Press, 1995).

27. J. Paul Heineman, Deposition, January 5, 1987 (USHMM: RG 09.008); Ed Granland, Deposition, October 7, 1986 (USHMM: RG 09.008).

28. Margaret Bourke-White, *Dear Fatherland, Rest Quietly: A Report on the Collapse of Hitler's "Thousand Years"* (New York: Simon & Schuster, 1946), 73.

29. "Report on Inspection of German Concentration Camp at Buchenwald," in *Monthly Intelligence Summary for Psychological Warfare: How Much Do the Germans Know?*, April 30, 1945.

30. Blumenson, *Patton Papers,* 692.

31. Pulitzer, *Report to the American People,* 104.

32. "Seiller Tells of Finding Nazi Atrocity Camps," *Lousville Courier Journal,* June 1946 (USHMM: RG 09.008).

33. Lawrence Douglas, "The Shrunken Head of Buchenwald: Icons of Atrocity at Nuremberg," *Representations* (Summer 1998), 39–64.

34. "Opfer, Tat, Aufstieg: Vom Konzentrationslager Buchenwald zur Nationalen Mahn—und Gedenkstätte der DDR," in *Versteinertes Gedenken: Das Buchenwald Mahnmal von 1958* (Buchenwald: Stiftung Gedenkstatten Buchenwald und Mittelbau Dora, 1997).

35. J. Paul Heineman, Deposition, January 5, 1987 (USHMM: RG 09.008); Stephen E. Ambrose, *Citizen Soldiers: The U.S. Army from the Normandy Beaches to the Bulge to the Surrender of Germany* (New York: Simon & Schuster, 1997), 462; Malcolm Vending, *War Diary,* May 23, 1945 (Dachau Museum and Memorial Archives).

36. Harry Cain, "Burial for Bodies Removed from Wöbellin Concentration Camp," May 9, 1945 (USHMM: RG 04.044).

37. Julien D. Saks, Deposition (USHMM: RG 09.008).

38. *Atrocities: A Study of German Reactions,* Supreme Headquarters Allied Expeditionary Force, Psychological Warfare Division, Intelligence Section, June 21, 1945, 1; *KZ: Bildbericht aus fünf Konzentrationslagern,* Amerikanischen Kriegsinformationsamt im Auftrag des Oberbefehlshabers der Allierten Streitkräfte, Spring 1945, 1.

39. *Atrocities,* 2–3.

40. Ibid, 5, 6.

41. Ibid, 5, 7; *Amtliches Material zum Massenmord von Katyn* (Berlin: Gedruckt

im Deutschen Verlag, 1943); David Welch, *The Third Reich: Politics and Propaganda* (London: Routledge, 1995), 112.

42. Joseph Stalin, February 23, 1942 speech, quoted in Pike, *Politics of Culture*, 12. See also Norman M. Naimark, *The Russians in Germany: A History of the Soviet Zone of Occupation, 1945–1949* (Cambridge: Harvard University Press, 1995), and David Pike, *The Politics of Culture in Soviet-Occupied Germany, 1945–1949* (Stanford: Stanford University Press, 1992).

43. *Atrocities*, 7.

44. Milton, "Confronting Atrocities," 60–61; Knigge, "Opfer, Tat, Aufstieg," 11–13.

45. Byron Price, "Relations between the American Forces of Occupation and the German People," report to President Truman, in *Department of State Bulletin* 13, no. 336 (December 2, 1945): 891. In December 1941, President Roosevelt appointed Price director of the Office of Censorship, a post he held until the office's closure on August 14, 1945. President Truman then appointed Price as his representative to investigate relations between American occupation forces and the German people. Price also served as an adviser to Generals Dwight D. Eisenhower and Lucius D. Clay, from September to November 1945. See Michael S. Sweeney, *Secrets of Victory: The Office of Censorship and the American Press and Radio in World War II* (Chapel Hill: University of North Carolina Press, 2001).

46. Joseph W. Bendersky, *The "Jewish Threat": Anti-Semitic Politics of the U.S. Army* (New York: Basic Books, 2000), 227–348; Harold Hurwitz, interview with the author, Berlin, 1998.

47. Rebecca Boehling, *A Question of Priorities: Democratic Reforms and Economic Recovery in Postwar Germany* (New York: Berghahn Books, 1996), 51.

48. Elizabeth Heineman, "Memories of Germany's 'Crisis Years' and West German National Identity," in Hanna Schlisser, ed., *The Miracle Years: A Cultural History of West Germany, 1946–1968* (Princeton: Princeton University Press, 2001), 38. Also see Petra Goedde, "From Villains to Victims: Fraternization and the Feminization of Germany, 1945–1947," *Diplomatic History* 23, no. 1 (Winter 1999): 1–48.

49. "What the American Soldier in Germany Says about Germany and the Germans," based on a questionnaire survey of enlisted men stationed in Germany, September 1945 (Landesarchiv Berlin: OMGBS, RG 260, box 8–2, folder 3).

50. Cited in Lawrence Douglas, "Film as Witness: Screening *Nazi Concentration Camps* before the Nuremberg Tribunal," *Yale Law Journal* 105: 449–75 (November 1995), 461.

51. *Occupation* (U. S. Forces, European Theater, 1946), 20.

52. Ibid., 19.

53. William J. Caldwell, "Democracy Comes to Dachau," *HICOG Monthly Bulletin*, May 1951, 23.

54. Ibid., 24.

55. "German Attitudes on Collective Guilt," *Information Control Review*, April 26, 1947.

56. David Crew, "Remembering German Pasts: Memory in German History, 1871–1989," *Central European History* 33, no.2 (2000): 218.

Chapter Two

1. Gerald M. Mayer, "American Motion Pictures in World Trade," *Annals of the American Academy of Political and Social Science* 254 (November 1947): 34. See also Lucius D. Clay, *Decision in Germany* (Melbourne: William Heinemann, 1950), 282.

2. Harold Zink, *The United States in Germany, 1944–1955* (Princeton: D. Van Nostrand, 1957), 234.

3. See Paul Rotha, "Some Principles of Documentary," in Richard Meran Barsam, ed., *Nonfiction Film: Theory and Criticism* (New York: E. P. Dutton, 1976), 42–55; Astrid Böger, *People's Lives Public Images: The New Deal Documentary Aesthetic* (Tübingen: Gunter Narr Verlag, 2001); and Jack C. Ellis and Betsy A. McLane, *A New History of Documentary Film* (New York: Continuum, 2005).

4. Stephen Tallents, "The Documentary Film," in Barsam, *Nonfiction Film,* 56, 57. For Grierson's definition, see Richard Meran Barsam, "The Nonfiction Film Idea," also in Barsam, *Nonfiction Film,* 15.

5. Lorentz (1905–1995) began his career as a journalist and film critic. In 1930 he wrote, with Morris L. Ernst, *Censored: The Private Life of the Movies,* which exposed the complex network of film censorship in the United States and criticized the structure of American media. In 1934 he published *The Roosevelt Year,* a collection of photographs and text. In 1942 President Roosevelt commissioned Lorentz as a major of the U.S. Army Air Corps. Lorenz graduated from the U.S. Army Air Force Officers Training School in January 1943 and worked closely with John Steinbeck, who had also been made a major in the Air Corps. Lorentz later commanded the Overseas Technical Unit of the Air Transport Command (ATC) and supervised the production of the ATC pilot training films.

6. See Gregory D. Black, *The Catholic Crusade Against the Movies, 1940–1975* (Cambridge: Cambridge University Press, 1997), and Richard S. Randall, *Censorship of the Movies: The Social and Political Control of a Mass Medium* (Madison: University of Wisconsin Press, 1968). In May 1945 the Hays Office turned down a seal for the American documentary *Atrocities* because of the word "damn" and the gruesome material shown. "Nazi Atrocity Films Real Shockers but U.S. Audiences Take It; Some Cuts," *Variety,* May 9, 1945, 18.

7. Robert T. Elson, "Times Marches on the Screen," in Barsam, *Nonfiction Film,* 93. See Ellis and McLane, *New History of Doumentary Film,* 78–80, and Raymond Fielding, *The March of Time, 1935–1951* (New York: Oxford University Press, 1978), 185.

8. See Thomas Doherty, *Projections of War: Hollywood, American Culture, and World War II* (New York: Columbia University Press, 1993); Russel Earl Shain, "An Analysis of Motion Pictures about War Released by the American Film Industry, 1939–1970," PhD thesis, University of Illinois, Urbana, 1971; and Clayton R. Koppes

and Gregory D. Black, *Hollywood Goes to War: How Politics, Profits, and Propaganda Shaped World War II Movies* (New York: Basic Books, 1987).

9. See Doherty, *Projections of War,* 70–75; Joseph McBride, *Frank Capra: The Catastrophe of Success* (New York: Simon & Schuster, 1992); William Alexander, *Film on the Left: American Documentary Film From 1931 to 1942* (Princeton: Princeton University Press, 1981); Erik Barnouw, *Documentary: A History of the Non-fiction Film* (New York: Oxford University Press, 1993); and Barsam, *Nonfiction Film.*

10. Edward L. Munson Jr., "The Army Pictorial Service," *Business Screen* 7, no. 1 (January 1948): 34; Klaus Kämpfe, "Educating the Re-Educators: Films in the Morale-Program of the U.S. Army," in Michael Hoenisch, Klaus Kämpfe, and Karl-Heinz Pütz, eds., *USA und Deutschland: Amerikanische Kulturpolitik 1942–1949,* no. 15 (Berlin: J. F. Kennedy–Institut für Nordamerikastudien, 1980), 160.

11. Emanuel Cohen, "Film Is a Weapon," *Business Screen* 7, no. 1 (January 1948): 74.

12. R. C. Barrett, "The Signal Corps Photographic Center," *Business Screen,* Issue 1, Vol. 7, January 1948, 37; Karl Heniz Pütz, "Business or Propaganda? American Films and Germany, 1942–1946," in Ekkehart Krippendorff, ed., *The Role of the United States in the Reconstruction of Italy and West Germany, 1943–1949* (Berlin: J. F. Kennedy–Institut für Nordamerikastudien, 1981), 188.

13. Examples include Josef von Sternberg (dir.), *The Town* (1944); Alexander Hammid, *Vally of the Tennessee* (1944) and *A Better Tomorrow* (1945); Helen Grayson and Larry Madison, *The Cummington Story* (1945), Willard Van Dyke's *Oswego* (1943), *Steeltown* (1944), *Pacific Northwest* (1944), *San Francisco* (1945), and *Northwest, U.S.A.* (1945); Jules Bucher, *The Window Cleaner* (1945); Henwar Rodakiewicz, *Capital Story* (1944); and John Houseman, *Tuesday in November* (1945).

14. John Grierson, "The Nature of Propaganda,"in Barsam, *Nonfiction Film,* 31.

15. As late as 1939, 40 percent of Hollywood revenues came from abroad, and thirty-five thousand theaters in Europe showed American films. Hollywood made great efforts to ensure that distribution continued unhindered by the European conflict, acquiescing, for example, to the Nazi race laws that banned Jewish actors from appearing on the German screen. Koppes and Black, *Hollywood Goes to War,* 21–34.

16. Richard Grunberger, *A Social History of the Third Reich* (London: Penguin Books, 1971), 476.

17. David Welch, *The Third Reich: Politics and Propaganda* (London: Routledge, 1995), 46. For analysis of the Nazi film industry, see Linda Schulte-Sasse, *Entertaining the Third Reich: Illusions of Wholeness in Nazi Cinema* (Durham: Duke University Press, 1996); David Welch, *Propaganda and the German Cinema* (Oxford: Clarendon Press, 1983); Eric Rentschler, *Ministry of Illusion: Nazi Cinema and Its Afterlife* (Harvard: Harvard University Press, 1996); Siegfried Kracauer, "Propaganda and the Nazi War Film," in *From Caligari to Hitler: A Psychological History of the German Film* (Princeton: Princeton University Press, 1947); and Hilmar Hoff-

mann, *The Triumph of Propaganda: Film and National Socialism, 1933–1945* (Providence: Breghahn Books, 1996). For analysis of "killing films," see Michael Burleigh, *Ethics and Extermination: Reflections on Nazi Genocide* (Cambridge: Cambridge University Press, 1997).

18. Kracauer, "Propaganda and the Nazi War Film," 275.

19. Susan Tegel, "Veit Harlan and the Origins of 'Jud Süss,' 1938–1939: Opportunism in the Creation of Nazi and Anti-Semitic Film Propaganda," *Historical Journal of Film, Radio, and Television* 16, no. 4 (1996): 515–16.

20. See Paul Virilio, *Guerre et cinéma: Logistique de la perception* (Paris: Cahiers du Cinema, 1991), and Christian Delage, *La vision nazie de l'histoire a travers le cinéma documentaire du Troisième Reich* (Laussane: Editions L'Age du Homme, 1989).

21. Joachim Joesten, "The German Film Industry, 1945–1948," *New Germany Report*, no. 3, May 1948, 3. See also "Report from Germany: U.S. Films in Reopened Theaters," *Film News* 7, no. 3 (November–December 1945): 20; Robert R. Shandley, *Rubble Films: German Cinema in the Shadow of the Third Reich* (Philadelphia: Temple University Press, 2001); Heide Fehrenbach, *Cinema in Democratizing Germany: Reconstructing National Identity after Hitler* (Chapel Hill: University of North Carolina Press, 1995); and Daniela Berghahn, *Hollywood behind the Wall: The Cinema of East Germany* (Manchester: Manchester University Press, 2005).

22. OMGBS ISD, "Cinema, Theater and Music," July 12, 1945 (U.S. National Archives [NARA]: RG 260, Box 8-2, Folder 1, 1).

23. Frederic Ullman, "Russia Eclipses U.S. on Pix in Reich," *Variety*, 1947, 1.

24. Gladwin Hill, "Our Film Program in Germany: II. How Far Was It a Failure," *Hollywood Quarterly* 2:131–37 (1947), 134.

25. Roger Smither, "*Welt im Film:* Anglo-American Newsreel Policy," in Nicholas Pronay and Keith Wilson, eds., *The Political Re-Education of Germany and Her Allies after World War II* (London: Croom Helm, 1985); Edward C. Breitenkamp, *The U.S. Information Control Division and Its Effect on German Publishers and Writers 1945 to 1949* (Grand Forks, ND: University Station, 1953).

26. Koppes and Black, *Hollywood Goes to War*, 140; Martin Quigley, "Importance of the Entertainment Film," *Annals of the American Academy of Political and Social Science* 254 (November 1947): 68; and Robert Joseph, "Cinema," *Arts and Architecture* 63 (December 1946): 14.

27. Brigitte J. Hahn, *Umerziehung durch Dokumentarfilm? Ein Instrument amerikanischer Kulturpolitik im Nachkriegsdeutschland (1945–1953)* (Munster: Lit Verlag, 1997), 1.

28. Robert Joseph, "Film Program for Germany," *Arts and Architecture* 62 (July 1945): 16.

29. Johaness Hauser, *Neuaufbau der westdeutschen Filmwirtschaft 1945–1955* (Freiburg: Centaurus-Verlagsgesellschaft, 1989), 193, 201; Charles A. H. Thomson, *Overseas Information Service of the United States Government* (Washington, DC: Brookings Institution, 1948), 41, 223.

30. "Selection of Motion Pictures Shown to German Youth," September 20, 1946 (NARA: RG 260, Box 24, Folder 5). See Robert A. McClure, "Selection of American Films for Germany," February 9, 1946 (NARA: RG 260, Box 262, Folder 12); Robert Joseph, "Confession and Reaffirmation," *Arts and Architecture* 63 (May 1946): 14.

31. Robert A. McClure, "Selection of American Films for Germany," February 9, 1946 (NARA: RG 260, Box 262, Folder 12).

32. Robert Joseph, "Our Film Program in Germany," *Hollywood Quarterly* 2, no. 2 (January 1947): 122–30.

33. Robert Joseph, "Comment and Criticism," *Arts and Architecture* 62 (September 1946): 16; Carol H. Denison, "First Things First," *Film News* 9, no. 5 (January 1949): 13.

34. Henry P. Pilgert, *The History of the Development of Information Services through Information Centers and Documentary Films* (Bonn: Historical Division Office of the Executive Secretary Office of the U.S. High Commissioner for Germany, 1951), 63.

35. Haynes R. Mahoney, "Germany Goes to the Movies," *Monthly Information Bulletin HICOG,* January 1951, 16.

36. Gladwin Hill, "Our Film Program in Germany: II. How Far Was It a Failure," *Hollywood Quarterly* 2:131–37 (1947), 135.

37. Paul M. A. Linebarger, *Psychological Warfare* (Washington, DC: Combat Forces Press, 1948), 210.

38. David Culbert, "American Film Policy in the Re-education of Germany after 1945," in Pronay and Wilson, *Political Re-Education of Germany,* 176.

39. Nicolas Lasson and Annette Michelson, "Notes of the Images of the Camps," *October* 90 (Autumn 1999), 26. See also Warren Wade, "How Is Combat Footage Used?," *Business Screen* 7, no. 1 (January 1948): 82. International organizations working with displaced persons used the material to assemble several films intended for fund-raising outside Germany, such as *These Are the People (*1946), *Report on the Living* (1947), *Placing the Displaced* (1948), and *Lang is der Weg* (1948).

40. Elmer Davis, quoted in "Films, Radio to Bombard Nazis," *Variety,* April 24, 1945.

41. Brewster Chamberlain, "Death Mills," in George O. Kent, ed., *Historians and Archivists: Essays in Modern German History and Archival Policy* (Fairfax, VA: George Mason University Press, 1991), 234, 235. Bernstein proposed an Anglo-American movie on the liberation of Belsen, detailing three versions of the film—one for Germans in Germany, one for German prisoners of war, and one for audiences from neutral, liberated, and allied territories—in a document dated April 30, 1945. All three versions would discuss the victims, the perpetrators, and the conditions of the camps. On May 25 Davidson Taylor responded with a memo explaining that showing atrocity material to a German audience was counterproductive, and on July 9 William D. Patterson, chief of the Film Division of OWI, informed Bernstein that

OWI would provide neither personnel nor facilities for a full-length atrocity film. Interestingly, the rejection of atrocity material as a method of reeducation and the directive to create *Todesmühlen* came at the same time (May 1945) and from the same place (PWD). Elizabeth Sussex, "The Fate of F3080," *Sight & Sound* 53, no. 2 (Spring 1984): 92–105.

42. See *International Biographical Dictionary of Central European Emigrés, 1933–1945,* vol. 1 (Munich: K. G. Saur, 1983); Rudolf Lang, "Hanuš Burger 75 Jahre," in *Studienkreis Rundfunk und Geschichte. Mitteilungen,* Jg. 10, no. 3 (July 1, 1984); and Hanus Burger, *Der Fruhling war es wert: Erinnerung* (Munich: C. Bertelsman Verlag, 1977).

43. Chamberlain, "Death Mills," 237; Michael Hoenisch "Film as an Instrument of the U.S. Reeducation Program in Germany after 1945 and the example of *Todesmühlen,*" in Krippendorff, *Role of the United States,* 138–39.

44. Ralph Willett, "Billy Wilder's 'A Foreign Affair' (1945–1948): The Trials and Tribulations of Berlin," *Historical Journal of Film, Radio, and Television* 7, no. 1 (1987): 3.

45. Burger, *Der Fruhling war es wert,* 255.

46. Ibid., 254–60; Hoenisch, "Film as an Instrument," 137–40; Chamberlain, "Death Mills," 237–38; David Culbert, "American Film Policy in the Re-education of Germany after 1945," in Pronay and Wilson, *Political Re-Education of Germany,* 177–78.

47. Charlotte Chandler, *Nobody's Perfect, Billy Wilder* (New York: Schuster & Schuster, 2002); Maurice Zolotow, *Billy Wilder in Hollywood* (Pompton Plains: Limelight Editions, 2004).

48. Cameron Crowe, *Conversations with Wilder* (New York: Alfred A. Knopf, 1999), 21.

49. Decades after the war, Wilder told Cameron Crowe that he very much wished he had been able to direct *Schindler's List.* Ibid., 20–21.

50. On the Jewish-American response to the Holocaust, see D. G. Myers, "Jews without Memory: Sophie's Choice and the Ideology of Liberal Anti-Judaism," *American Jewish History* 88, no. 1 (March 2000): 499–529; Laurel Leff, "A Tragic 'Fight in the Family': The New York Times, Reform Judaism and the Holocaust," *American Jewish History* 88, no. 1 (March 2000): 3–51; Rafael Medoff, "'Our Leaders Cannot Be Moved': A Zionist Emissary's Reports on American Jewish Responses to the Holocaust in the Summer of 1943," *American Jewish History* 88, no. 1 (March 2000): 115–26; and Edward Alexander, *The Resonance of Dust: Essays on Holocaust Literature and Jewish Fate* (Columbus: Ohio State University Press, 1979).

51. Eric T. Clarke, Chief, Film, Theater and Music Control, memorandum to Brigadier General Robert A. McClure, January 9, 1946 (NARA: RG 260, Box 262).

52. J. H. Hills, Deputy Director of the Division, memo to Chief, Information Control Branch, "Compulsory showing of atrocity film," February 18, 1946 (NARA: RG260, Box 262). See Edward T. Peeples, Executive Officer, memo to Lt. Colonel Ir-

ving Dilliard, OMGUS Bavaria, "Attendance at atrocity film showings," February 8, 1946 (NARA: RG260, Box 262).

53. OMGUS recorded German attendance at films and conducted polls to determine viewer reactions. These activities were carried out by the Reorientation Branch of the Civil Affairs Division of the War Department, by the Office of International Information and Cultural Affairs of the State Department, and by OMGUS. G. H. Garde, Lieutenant Colonel AGD, Adjunct General, for the Acting Deputy Military Governor, September 20, 1946 (NARA: RG 260, Box 24, Folder 5); Robert Joseph, "The Germans See Their Concentration Camps," *Arts and Architecture* 63 (September 1946): 14.

54. Gerhard Kreische caricature, *Ulenspiegel,* September 1, 1946.

55. "German Reactions: Mills of Death," *Military Government Weekly Information Bulletin,* no. 30, February 23, 1946; "The Atrocity Film in Berlin," U.S. Headquarters Berlin District, OMGUS, Information Services Control Section, Intelligence Sub-Section, Public Opinion Surveys Unit-Berlin Area, April 9, 1946 (NARA: RG 260, box 8-3, folder 3).

56. "Weekly cable to Washington," OMGBS Information Services Control section, March 29, 1946 (Landesarchiv Berlin: RG 260, box 9-1, folder 38).

57. Michael Josselson, Chief of Intelligence, "Request for Special Report on German Reactions to American Films," September 4, 1946 (Landesarchiv Berlin: RG 260, 4/8–2/8).

58. "German Reactions: Mills of Death," 14.

59. "Military Government Information Control," *Monthly Report of Military Governor U.S. Zone,* January 1946, 10.

60. "The World in Film," *OMGUS Weekly Information Bulletin,* Number 70/2, December 1946.

61. On November 29, 1946, ICD issued its first ten licenses to German film producers, and production resumed in the spring of 1947. Thomas R. Hutton, Chief ISB, OMGBS, Final Report Information Services Branch, July 17, 1949 (Landesarchiv Berlin: OMGBS, RG 260, box 12-1, folder 31). See Hauser, *Neuaufbau der westdeutschen Filmwirtschaft,* and Adolf Heinzlmeier, *Nachkriegsfilm und Nazifilm* (Frankfurt am Main: Oase Verlag, 1988).

62. Hahn, *Umerziehung durch Dokumentarfilm?,* 272. DFU films include *Achtung, Mücken!* ("Mosquitoes!," 1947), *Es liegt an Dir* ("It's Up to You," 1947), *Heimat im Moor* ("Home on the Moor," 1948), *Hunger* (1948), *Ich und Mr Marshall* ("Me and Mr. Marshall," 1948), *Ergebnis: Positiv* ("Reaction: Positive," 1948), *Bauern helfen Sich Selbst* ("Rural Coop," 1948), *Schritt für Schritt* ("Step by Step," 1948), *Nürnberg und seine Lehren* ("Nuremberg Trial," 1948), *Zwischen Ost und West* ("Between East and West," 1948), *Die Brücke* ("The Bridge," 1949), *Made in Germany* (1949), *Marschieren, marschieren!* ("Marching, Marching!," 1949), *Weiss/Gelb/Schwarz* ("White, Yellow, Black," 1949), and *Zwei Städte* ("Two Cities," 1949).

63. Stuart Schulberg, "Of All People," Hollywood Quarterly 4, no. 2 (Winter 1949): 206–8.

64. Sandra Schulberg and Ed Carter, "Selling Democracy: Films of the Marshall Plan, 1948–1953," catalog for national tour, 2005–2006.

65. Military Government of Germany, "Monthly Report of Military Governor U.S. Zone," *Information Control,* December 1947, 12.

66. See Alon Confino, "Remembering the Second World War, 1945–1965: Narratives of Victimhood and Genocide," *Cultural Analysis* 4 (2005): 47–75; Aleida Assmann, "On the (In)Compatibility of Guilt and Suffering in German Memory," *German Life and Letters* 59, no. 2 (April 2006): 187–200; and Robert G. Moeller, *War Stories: The Search for a Usable Past in the Federal Republic of Germany* (Berkeley: University of California Press, 2001).

67. Stuart Schulberg, letter to Sone and Ben, May 7, 1948. "Sone" is Sonya Schulberg O'Sullivan, Stuart's elder sister, and "Ben" was Benjamin O'Sullivan, her husband. The letter is in the possession of Sandra, K. C., Peter, and Jon Schulberg, the heirs of Barbara and Stuart Schulberg, and was kindly provided by Sandra Schulberg.

68. Ibid.

69. Barbara Goodrich Schulberg, letter to parents (Carter Lyman Goodrich and Florence Perry Nielsen), March 29, 1948. Stuart Schulberg, letter to Sone and Ben, May 7, 1948. Stuart Schulberg, personal papers; provided by Sandra Schulberg.

70. See John Willoughby, *Remaking the Conquering Heroes: The Social and Geopolitical Impact of the Post-War American Occupation of Germany* (New York: Palgrave, 2001), chap. 3, "The Corrosive Racial Divide."

71. Brenda Gayle Plummer, ed., *Window on Freedom: Race, Civil Rights, and Foreign Affairs, 1945–1948* (Chapel Hill: The University of North Carolina Press, 2003), 27. The evolution of postwar anti-American propaganda in *L'Humanité,* the official newspaper of the French Communist Party, is illustrative. From 1945 to late 1946 references to America were few and bland, and the visits of Hollywood figures such as Rita Hayworth were covered extensively. Hollywood represented American glamour and the promise of a new beginning. With the emergence of the Cold War, the French Communists initiated an anti-American campaign focused on (a) denunciation of American financial and economic imperialism, (b) prediction of an imminent financial, economic, and social crisis in the United States, (c) assertion of the risk of cultural "Americanization," and (d) depiction of the United States as a racist country. The Communist agenda of *autodéfense culturelle* sought to "protect" French traditions from U.S. encroachment. American culture and social habits were broadly denounced, as was the Marshall Plan. The American film industry was a particular target; starting in 1947, enthusiasm for Hollywood was replaced with rhetoric on the need to defend the French film industry. The issue of race played an important role in the way *L'Humanité* depicted American society in 1947. There were a few positive stories about the United States that year: coverage of former vice president Henry A. Wallace, the Progressive Party candidate for president in the 1948 elections; of the participation of Dr. Gene Weltfish in an international women's conference; and of well-known African American entertainers and athletes—Paul Robeson, Joe Louis,

and Josephine Baker. The rest of the paper's news coverage of the United States was negative, depicting the contradictions and dreariness of life in America, announcing impeding economic crises, and reporting strikes and antiblack violence. There were articles on lynching, on a racially motivated mob killing, and on the execution of two black boys. The title of a May 14 piece, "Truman's America is the Country of Triumphant Racism" exemplifies the tone of the articles. In 1948 the coverage of racial problems increased, with fifteen articles on American racism.

72. "29 Unions Discriminate," July 11, 1942; Vassar: Ruth Benedict papers, 53.7.

73. Teleconference request by OMGUS for "Brotherhood of Man,"August 30, 1946 to October 4, 1946, Special File of the Records Section of the Civil Affairs Division, War Department Special Staff, undated (NARA: RG 165, Box 254, File 62.2); Colonel E. E. Hume, memo, August 27, 1947, Special File of the Records Section of the Civil Affairs Division, War Department Special Staff (NARA: RG 165, Box 254, File 62).

74. See George W. Stocking Jr., ed., *A Franz Boas Reader: The Shaping of American Anthropology, 1883–1911* (Chicago: University of Chicago Press, 1974), and Elazar Barkan, *The Retreat of Scientific Racism: Changing Concepts of Race in Britain and the United States between the World Wars* (New York: Cambridge University Press, 1992).

75. Ruth Benedict and Gene Weltfish, *The Races of Mankind* (New York: Public Affairs Committee, Pamphlet No 85, 1943). The pamphlet was prepared under the supervision of the American Association of Scientific Workers, financed by a grant from the Alfred P. Sloan Foundation, and edited by the Public Affairs Committee (Pamphlet N° 85).

76. Colonel E. E. Hume, memorandum, August 27, 1947, Special File of the Records Section of the Civil Affairs Division, War Department Special Staff (NARA: RG 165, Box 254, File 62).

77. "Confidential memo on the Races of Mankind," Special File of the Records Section of the Civil Affairs Division, War Department Special Staff, undated (NARA: RG 165, Box 254, File 62.2). See also Ruth Benedict and Gene Weltfish, *Race: Science and Politics* (Wesport: Grenwood Press, 1940), 167–68.

78. Congressional Record—Senate, May 12, 1944, 4495–4500 (Vassar College Archives: Ruth Benedict papers, 53.1). In the same speech, Representative Bilbo attacked Boas—referred to as a member of the "Jewish race" by all the members of Congress involved in the debate—for "teachings that advocated the intermarriage of the races." Bilbo was a strong supporter of President Roosevelt's New Deal programs, yet championed a full gamut of racist and anti-Semitic causes. He tried to preserve the poll tax in southern states to prevent African Americans from voting, introduced a bill to solve "the Negro question" by relocating the "American Negro in a territory separate from the Whites" (specifically Liberia), and introduced a bill to protect the "integrity of the White race" by banning intermarriage in Washington, D.C. In 1946 a Senate committee found Bilbo guilty of accepting bribes from military contractors during the war; he was charged with corruption and denied his seat in 1947. Ad-

win Wigfall Green, *The Man Bilbo* (Baton Rouge: Louisiana State University Press, 1963); Chester Morgan, *Redneck Liberal: Theodore G. Bilbo and the New Deal* (Baton Rouge: Louisiana State University Press, 1985).

79. "Confidential memo on the Races of Mankind," Special File of the Records Section of the Civil Affairs Division, War Department Special Staff, undated (NARA: RG 165, Box 254, File 62.2). The suppression by the USO of *The Races of Mankind* prompted heated exchanges in the press. For Chester Barnard's defense of his decision, see "The U.S.O. and Races of Mankind," *Saturday Review of Literature* 27, no. 29 (July 15, 1944): 13.

80. Leonard Maltin, *Of Mice and Magic: A History of American Animated Cartoons* (New York: Plume, 1987), 327.

81. Hubley was the supervising director of United Productions of America (UPA), a small animation studio that made the Oscar-winning cartoon *Gerald McBoing Boing*. Most UPA artists were New Dealers, who believed that animation had a political dimension. In 1944 they made *Hell-Bent for Election,* a pro-Roosevelt short that was the first animated film to acknowledge segregation in America. UPA never produced another cartoon that promoted racial equality, but its subsequent films did not contain African American stereotypes. See Christopher P. Lehman, "The New Black Animated Images of 1946," *Journal of Popular Film and Television* 29, no. 2 (Summer 2001): 74–81; Maltin, *Of Mice and Magic;* and Charles Solomon, *The History of Animation: Enchanted Drawings* (New York: Wings Books, 1994).

82. Colonel W. E. Crist. memo to Director of Intelligence, War Department General Staff, May 20, 1947, Special File of the Records Section of the Civil Affairs Division, War Department Special Staff (NARA: RG 165, Box 254, File 62); cable dispatch, May 20, 1947, Special File of the Records Section of the Civil Affairs Division, War Department Special Staff (NARA: RG 165, Box 254, File 62); cover sheet, Special File: "Brotherhood of Man" (1946–1949) (NARA: RG165, CAD 062.2, Box 254, Entry 463).

83. Walter P. Reuther, letter to President Truman, August 15, 1947, Special File: "Brotherhood of Man" (1946–1949) (NARA: RG165, CAD 062.2, Box 254, Entry 463).

84. Kenneth C. Royall, letter to President Truman, September 2, 1947, Special File of the Records Section of the Civil Affairs Division, War Department Special Staff (NARA: RG 165, Box 254, File 62).

85. Richard M. Dalfiume, *Desegregation of the U.S. Armed Forces* (Columbia: University of Missouri Press, 1969), 139.

86. Raymond Daniell, "What the Europeans Think of Us: Their Ideas about Us Are Distorted, but We Have Fumbled the Job of Explaining Our Aims to Them," *New York Times,* November 30, 1947, SM7.

87. Walter P. Reuther, letter to President Truman, September 17, 1948, Special File of the Records Section of the Civil Affairs Division, War Department Special Staff (NARA: RG 165, Box 254, File 62).

88. Colonel McMahon, "memo for the record," October 20, 1948, Special File of the

Records Section of the Civil Affairs Division, War Department Special Staff (NARA: RG 165, Box 254, File 62).

89. Ibid.

90. William D. Hassett, secretary to the president, letter to Walter P. Reuther, November 5, 1948 (Truman Library Archives: "The Brotherhood of Man" file).

91. During World War II, African American soldiers in Europe and the Pacific were mainly restricted to service and supply units. Segregation proved economically cumbersome because the duplication of facilities and services was onerous and inefficient. In September 1945 newly appointed secretary of war Robert P. Patterson convened a board led by Lieutenant General Alvan C. Gillem Jr., which formulated a policy directive suggesting that soldiers be evaluated on the basis of merit and ability rather than race. The Army closed ranks in opposition to the Gillem Report. General Noce, then acting chief of staff at the Army Ground Forces headquarters, claimed that integration was unfeasible. Secretary of the army Royall interpreted the Gillem Report as proposing a policy of equality of opportunity within a segregationist context, and the U.S. Army continued its discriminatory, segregationist, and racist agenda. On July 26, 1948, Truman signed Executive Order 9981, which established "equality of treatment and opportunity for all persons in the armed services without regard to race, color, religion or national origin." The order also created the President's Committee on Equality of Treatment and Opportunity in the Armed Services, known as the Fahy Committee. The Fahy Committee eventually got the Army to open all units to qualified African American soldiers, with assignments made on the basis of individual ability. By the end of the Korean War in 1953, the U.S. military was almost completely desegregated. See Mary L. Dudziak, *Cold War Civil Rights: Race and the Image of American Democracy* (Princeton: Princeton University Press, 2000); Thomas Borstelmann, *The Cold War and the Color Line: American Race Relations in the Global Arena* (Cambridge: Harvard University Press, 2001); Morris J. MacGregor Jr., *Integration of the Armed Forces 1940–1965* (Washington, DC: Center of Military History, United States Army, 1985).

92. Karl F. Cohen, *Forbidden Animation: Censored Cartoons and Blacklisted Animators in America* (Jefferson: Mc Farland & Company, 1997), 176–77; see the *Fourth Report, Un-American Activities in California, 1948, Communist Front Organizations* (Sacramento: California, 1948).

Chapter Three

1. Prolog, *A Report on German Museums* (Berlin, Fall 1947); Edouard Roditi, "The Destruction of the Berlin Museums," *Magazine of Art* 42, no. 8 (December 1949), 306–12; Andre Kormendi, "What German Art Survives," *Art News,* October 1948, 41–43. For more on the art scene in occupied Germany, see Eckart Gillen and Diether Schmidt, eds., *Zone 5: Kunst in der Viersektorenstadt 1945–1951* (Berlin: Dirk Nishen, 1989); Yule F. Heibel, *Reconstructing the Subject: Modernist Painting in Western Germany, 1945–1950* (Princeton: Princeton University Press, 1995); Markus

Krause, *Galerie Gerd Rosen: Die Avantgarde in Berlin 1945–1949* (Berlin: Ars Nicolai, 1995); and Marion F. Deshmukh, "The Year Zero and Beyond," in George O. Kent, ed., *Historians and Archivists: Essays in Modern German History and Archival Policy* (Fairfax, VA: George Mason University Press, 1991).

2. James J. Sheehan, *Museums in the German Art World: From the End of the Old Regime to the Rise of Modernism* (Oxford: Oxford University Press, 2000), 98. See also Hans Belting, *The Germans and Their Art: A Troublesome Relationship* (New Haven: Yale University Press, 1998); Claudia Ruckert and Sven Kuhrau, eds., *Der Deutschen Kunst: Nationalgalerie und Nationale Identitat 1876–1998* (Amsterdam: Verlag der Kunst, 1998); and Werner Hofman, *Wie deutsch ist die deutsche Kunst?* (Leipzig: E. A. Seeman, 1999).

3. Belting, *Germans and Their Art*, 62.

4. Ibid., 52. Also see David Pan, "The Struggle for Myth in the Nazi Period," *South Atlantic Review* 65, no. 1 (Winter 2000): 41–57.

5. *Paris-Berlin, 1900–1933*. (Paris: Gallimard, 1992); Frederic Spotts, *Hitler and the Power of Aesthetics* (Woodstock: Overlook Press, 2002), 161. Also see Rose-Carol Washton-Long, *German Expressionism: Documents from the End of the Wilhelmine Empire to the Rise of National Socialism* (New York: G. K. Hall, 1993).

6. Milton Cohen, "Fatal Symbiosis: Modernism and WWI," *War, Literature, and the Arts* 8 (1996): 4. See Henry Grosshans, *Hitler and the Artists* (New York: Holmes & Meier, 1983), 48. The Verists were on the political left of the German avant-garde. Many of those on the right were painters of the New Objectivity (Neue Sachlichkeit) school, influenced by the metaphysical painting of Giorgio De Chirico. They emphasized classic forms in a static and atemporal context that invited interpretation and reflection. See Uwe Schneede, "Verisme et Nouvelle Objectivité," in *Paris-Berlin*, 238–61.

7. David Welch, *The Third Reich: Politics and Propaganda* (London: Routledge, 1993), 170–71.

8. Spotts, *Hitler and the Power of Aesthetics*, 9–27; Jonathan Petropoulos, *Art as Politics in the Third Reich* (Chapel Hill: University of North Carolina Press, 1996), 140.

9. For further discussion of the Nazi debates on German Expressionism, see Spotts, *Hitler and the Power of Aesthetics*.

10. Toby Clark, *Art and Propaganda in the Twentieth Century: The Political Image in the Age of Mass Culture* (New York: Harry N. Abrams, 1997), 61. See John Heskett, "Modernism and Archaism in Design in the Third Reich," in Brandon Taylor and Wilfried van der Will, eds., *The Nazification of Art: Art, Design, Music, Architecture, and Film in the Third Reich* (Winchester: Winchester Press, 1990).

11. The "Degenerate Art Exhibit" was seen by 2,009,899 visitors in Munich, before being sent to Berlin, Leipzig, Düsseldorf, and Hamburg. Spotts, *Hitler and the Power of Aesthetics*, 165.

12. Lincoln Kirstein, "Art in the Third Reich—Survey, 1945," *Magazine of Art* 38, no. 6 (October 1945): 231.

13. Peter Adam, *Art of the Third Reich* (New York: Harry N. Abrams, 1992), 117.

14. Spotts, *Hitler and the Power of Aesthetics,* 157.

15. The numbers are staggering if compared to contemporary figures. The most attended postwar show at the Haus der Kunst was the 1980 *Tutanchamun* exhibit, which attracted 652,700 visitors. *Haus der Kunst 1937–1997: Eine Historische Dokumentation* (Munich: Haus der Kunst München, 1997).

16. Steglitz Archive files. See also "Records of Property Division, December 6 1946 (NARA: RG 260, Box 84); Edith A. Standen, Daily Log August 1946—March 1947, Box 1 (National Gallery of Art archives [NGA]: Edith A. Standen papers, Box 1). Artists denied an OMGUS licence were not allowed to exhibit their work, teach at art schools, trade in art, or be art critics for newspapers or radio in the American zone and sector. The rigor of the exclusion process diminished with time. In October 1951, for instance, there was a large exhibit of work by painters and sculptors of Nazi fame at the Haus der Kunst, organized with the permission of the Bavarian Ministry of Culture. See Hellmut Lehmann-Haupt, "German Art Today," undated essay (MoMA: Lehmann-Haupt Papers, Box 4, File 16); letters from German artists to Lehmann-Haupt (MoMA: Lehmann-Haupt Papers, Box 7, File 1); Hellmut Lehmann-Haupt, "Behind the Iron Curtain," *Magazine of Art* 44 (1951): 87. Also see "Records of Property Division, December 6, 1946 (NARA: RG 260, Box 84); Edith A. Standen, Daily Log August 1946—March 1947 (NGA: Edith A. Standen papers, Box 1); Heinz Roemheld, memorandum, 15 October 1945 (Berlin Landesarchiv: OMGUS B, RG 260, 4/18–2/1); and Captain Gerard W. Van Loon, "The Theater Returns to the Bavarian Scene," *Military Government Weekly Information Bulletin,* no. 26 (26 January 1946), 5.

17. Marvin C. Ross, "SHAEF and the Protection of Monuments in Northwest Europe," *College Art Journal* 5, no. 2 (January 1946): 121. See Ralph W. Hammett, "Comzone and the Protection of Monuments in North-West Europe," *College Art Journal* 5, no. 2 (January 1946): 123–27; Robert K. Posey, "Protection of Cultural Materials during Combat," *College Art Journal* 5, no. 2 (January 1946): 127–31.

18. "The Beautiful Spoils: Monuments Men," *New Yorker,* March 8, 1947, 47. See Lynn H. Nicholas, *The Rape of Europa* (New York: Vintage, 1995), and Harry L. Cole and Albert K. Weinberg, "The Protection of Historical Monuments and Art Treasures," in *Civil Affairs: Soldiers Becoming Governors,* Special Studies Series on the United States Army in World War II (Washington, DC: Office of the Chief of Military History, Department of the Army, 1964).

19. Edith Appleton Standen, Memorandum to Craig Smyth, 13 November 1947 (NGA: Standen Papers, Box 1). Also see Hellmut Lehmann-Haupt, *Art under a Dictatorship* (New York: Oxford University Press, 1954), 198–99.

20. Edith Appleton Standen, "Report on Germany," *College Art Journal* 7, no. 3 (Spring 1948): 213.

21. "Decree of the Supreme Commander of Soviet Military Government and Supreme Commander of the Soviet Forces in Germany" (4 September 1945), quoted

in Erich Kuby, *The Russians and Berlin, 1945* (London: Heinemann 1965), 335–36. See also David Pike, *The Politics of Culture in Soviet-Occupied Germany, 1945–1949* (Stanford: Stanford University Press, 1992), and Bernard Genton, *Les alliés et la culture: Berlin, 1945–1946* (Paris: Presses Universitaires de France, 1988).

22. Pike, *Politics of Culture*, 97.

23. Ibid., 239, 226; Genton, *Les alliés et la culture*, 336–37.

24. Wolfgang Hutt, *Deutsche Malerei und Graphik im 20. Jahrhundert* (Berlin: Henschelverlag Kunst und Gesellschaft, 1969), 460.

25. See Pike, *Politics of Culture*, 238–42, 311, 536.

26. Modern art also challenged the American Regionalist school, which had aesthetic similarities to Soviet Socialist Realism and the Nazi New German Art but was considered an intrinsically American alternative to European avant-garde art. See James M. Dennis, *Renegade Regionalists: The Modern Independence of Grant Wood, Thomas Hart Benton, and John Curry* (Madison: University of Wisconsin Press, 1998) and Erika Doss, *Benton, Pollock, and the Politics of Modernism: from Regionalism to Abstract Expressionism* (Chicago: University of Chicago Press, 1991). For an account of postwar antimodernism in the United States, see Michael L. Krenn, *Fall-Out Shelter for the Human Spirit: American Art and the Cold War* (Chapel Hill: University of North Carolina Press, 2005).

27. William E. Daugherty, "Post-World War II Developments," in William E. Daugherty and Morris Janowitz, eds, *A Psychological Warfare Casebook* (Baltimore: Johns Hopkins University Press, 1958), 136.

28. J. Leroy Davidson, "Advancing American Art," and William Benton, "Advancing American Art," *Art News* 45, no. 8 (October 1946): 19; Edward Biberman, "The Attack on the American Artist," conference of the Hollywood Arts, Science, and Professions Council, July 9–13, 1947 (Hollywood, 1947). Also see Paul J. Braisted, ed., *Cultural Affairs and Foreign Relations* (Washington, DC: Columbia Books, 1968); Charles A. H. Thomson, *Overseas Information Service of the United States Government* (Washington, DC: Brookings Institution, 1948); Charles A. H. Thomson and Walter H. C. Laves, *Cultural Relations and U.S. Foreign Policy* (Bloomington: Indiana University Press, 1963); and Hans Tuch, *Communicating with the World: U.S. Diplomacy Overseas* (New York: St. Martin's Press, 1990).

29. "Memorandum on Art Program of the Office of International Information and Cultural Affairs," Department of State, February 1947 (Archives of American Art [AAA]: Advancing American Art File, MF #446); William Benton, memo, Department of State, February 10, 1947 (AAA: Advancing American Art, MF #445); *Catalog of 117 Oil and Water Color Originals by Leading American Artists*, 1947 (AAA: Advancing American Art, MF #403); Biberman "Attack on the American Artist"; Jo Gibbs, "State Department Art Classed as War Surplus," *Art Digest* 17, no. 9 (June 1, 1948): 9; Jo Gibbs, "State Department Shows 'Goodwill' Pictures," *Art Digest* 22, no. 13 (October 1, 1946): 13. For more recent analysis, see Jane de Hart Mathews, "Art and Politics in Cold War America," *American Historical Review* 81 (1976): 777; Serge Guilbaut, *How New York Stole the Idea of Modern Art: Abstract Expressionism,*

Freedom, and the Cold War (Chicago: University of Chicago Press, 1983); and Sigrid Ruby, "The Give and Take of American Painting in Postwar Western Europe" (presented at "The American Impact on Western Europe: Americanization and Westernization in Transatlantic Perspective," German Historical Institute, Washington, DC, March 25–27, 1999).

30. "'Government' Art Called Mediocre," *New York Times,* March 9, 1952; Taylor D. Littleton and Maltby Sykes, eds., *Advancing American Art* (Tuscaloosa: University of Alabama Press, 1989); Hugo Weisgall, *Advancing American Art 1947* (Prague: U.S. Information Service, 1947); "Exposing the Bunk of So-Called Modern Art," *New York Journal,* March 12, 1946.

31. Marilyn Robb, "Chicago," *Art News* 46, no. 11 (January 1948): 39.

32. Biberman, "Attack on the American Artist," 241; Sigrid Ruby, *Have We an American Art?* (Weimar: Verlag und Datenbank für Geisteswissenschaften, 1999), 72.

33. Guilbaut, *How New York Stole the Idea of Modern Art,* 4. Also see Frances Stonor Saunders, *The Cultural Cold War: The CIA and the World of Arts and Letters* (New York: New Press, 1999). In 1953 Dondero became the Chairman of the House Committee on Public Works, and continued to wage battle against art he considered subversive and un-American.

34. George Dondero quoted in Littleton and Sykes, *Advancing American Art,* 45.

35. Fred Busbey, quoted in *U.S. Congressional Record* 93 (May 13, 1947): 5221–22. Busbey, a militant anticommunist and member of the House Un-American Activities Committee, declared in 1943 that the foreign-language division of the Office of War Information was full of Communists and fellow travelers intent on undermining the American political system. See "Fred Busbey Dies," *New York Times,* February 13, 1966, 84.

36. John Taber, letter to George C. Marshall, Secretary of State, February 4, 1947 (AAA: Advancing American Art, #444).

37. "Questions concerning the Department's Art Program Asked by Edward Alden Jewell in the New York Times, June 15, 1947, with answers" (AAA: Advancing American Art, #500); Fred Othman, "Scrambled Egg Art Sale," *Washington Daily News,* May 14, 1948; "Senate Resolution No. 21" (AAA: Advancing American Art, #391); *Examiner* headline, quoted in Biberman, "Attack on the American Artist," 241.

38. Gibbs, "State Department Art Classed as War Surplus," 9; "Protests to State Department Mount," *Art News* 46, no. 4 (June 1947): 8.

39. "Disposal of 79 Oil Paintings at Present in the Possession of the Division of Libraries and Institutes," September 22, 1947 (AAA: Advancing American Art, # 527). See also Peyton Boswell, "When Art Becomes 'War Surplus,'" *Art Digest,* June 1, 1948, 7. Among the paintings sold below market value were two of Ben Shahn's most important works, *Hunger* and *Renascence,* then on loan to the Museum of Modern Art for an exhibit that ran from September 1947 to January 1948.

40. Gibbs, "State Department Art Classed as War Surplus"; Richard H. Heindel, letter to J. Leroy Davidson, April 1, 1947 (AAA: Advancing American Art); Robb, "Chicago"; Ruby, "Give and Take of American Painting."

41. Lehmann-Haupt, "Behind the Iron Curtain," 84. See Pike, *Politics of Culture*, 310, 530–36.

42. Lehmann-Haupt memorandum, October 3, 1946 (MoMA: Lehmann-Haupt Papers, Box 7, Folder 2).

43. Alfred Barr, director of New York's Museum of Modern Art, and Lincoln Kirstein, Lehmann-Haupt's future colleague at MFA&A, were also aware of the importance of the fine arts as an instrument for mass political education in the Third Reich. See Kirstein, "1945," and Alfred H. Barr Jr., "Art in the Third Reich—Preview 1933," *Magazine of Art* 38, no. 6 (October 1945): 212–22. In 1950, back from Germany, Lehmann-Haupt received a research grant from the Rockefeller Foundation to investigate the political use of the fine arts by totalitarian governments. The resulting book, *Art under Dictatorship* (1954), analyzed Nazi art policy thirty years before it became an issue of academic concern (Rockefeller Archive Center, RG 1.2, Series 200, Box 3a1).

44. Lehmann-Haupt, memorandum to Richard Howard and Charles M. Fleischner, November 15, 1947 (MoMA: Lehmann-Haupt Papers, Box 7, Folder 2 [1901–1988]). Fleischner, a lieutenant commander of the U.S. Naval Reserve, was deputy chief of the MFA&A in Berlin starting in 1946.

45. Lehmann-Haupt, memorandum to Howard, 12 February 1947 (MoMA: Lehmann-Haupt Papers, Box 7, Folder 2).

46. Lehmann-Haupt, "German Art Today," undated essay (MoMA: Lehmann-Haupt Papers, Box 4, Folder 16); Lehmann-Haupt, "Behind the Iron Curtain."

47. Lehmann-Haupt, "Art under Totalitarianism," Metropolitan Museum of Art, March 1952, 62.

48. Lehmann-Haupt, memorandum to Howard, 12 February 1947 (MoMA: Lehmann-Haupt Papers, Box 7, Folder 2).

49. See "Final Report on Field Trip to Bavaria by Civil Arts Administration Officer from January 3 to 26, 1947"; Lehmann-Haupt, memorandum to Howard, 20 March 1947; "Artists correspondence with Lehmann-Haupt" (all MoMA: Lehmann-Haupt Papers, Box 7, Folders 1 and 2); "Art in Postwar Berlin," undated essay (MoMA: Lehmann-Haupt Papers, Box 1, Folder 16); "Bavarian Reactions to Modern Art," *Information Control Review*, No. 18, 1946.

50. Lehmann-Haupt, memorandum to Howard, 12 February 1947, and Lehmann-Haupt, memorandum to Howard and Fleischner, 15 November 1947 (MoMA: Lehmann-Haupt Papers, Box 7, Folder 2).

51. Lehmann-Haupt, "German Art Today," undated essay (MoMA: Lehmann-Haupt Papers, Box 4, Folder 16).

52. Lehmann-Haupt, memorandum to Howard, March 20, 1947 (MoMA: Lehmann-Haupt Papers, Box 7, Folder 2).

53. Richard Howard, letter to Frederick A. Sweet, associate curator of painting and sculpture, Art Institute of Chicago, June 16, 1947 (NARA: RG 260, Box 216, Folder 4).

54. Lucius D. Clay, letter to Frederick A. Sweet, June 18, 1947 (MoMA: Lehmann-Haupt Papers, Box 7, Folder 2).

55. Edith Standen, letter to Craig Hugh Smyth, November 12, 1947 (NGA: Smyth Papers, Box 1).

56. Ibid.

57. Standen, memorandum to Smyth, November 13, 1947 (NGA: Standen Papers, Box 1).

58. Ibid., 2–4.

59. Howard, letter to Forest Huttenlocher, September 29, 1948 (Des Moines Art Center: Howard papers). See also Standen, *Monuments, Fine Arts, and Archives: May 1945—July 1947,* undated (NGA: Standen Papers, RG 280, Box 2), and records of the Museum, Fine Arts, and Archives Section (USHMM: RG 260).

Chapter Four

1. The Central Intelligence Agency (CIA), created in 1947, integrated a number of administrative units dedicated to espionage and clandestine activities; its covert anticommunism cultural programs began in the 1950s. See David Caute, *The Dancer Defects: The Struggle for Cultural Supremacy during the Cold War* (Oxford: Oxford University Press, 2005); Frances Stonor Saunders, *The Cultural Cold War: The CIA and the World of Arts and Letters* (New York: New Press, 1999); and Giles Scott-Smith and Hans Krabbendam, eds., *The Cultural Cold War in Western Europe 1945–1960* (London: Frank Cass, 2003). U.S. law defines covert action as activity meant "to influence political, economic, or military conditions abroad, where it is intended that the role of the United States Government will not be apparent or acknowledged publicly." While the action itself need not to be clandestine, its sponsor remains anonymous and can therefore deny responsibility. Jennifer D. Kibbe, "The Rise of the Shadow Warriors," *Foreign Affairs,* March/April 2004, 102–15.

2. Lutzeier, also active in the Friends of Contemporary Art, a similar group in Frankfurt, became coordinator for U.S. Information Centers in Hesse in 1950. Ursula Bluhm, memorandum, June 1, 1950 (AAA: Constable Papers, 610–12).

3. *Weekly Information Bulletin,* Office of the Assistant Chief of Staff, G-5 Division USFET, Information Branch, Number 117, November 1947, 9; Howard, memorandum to Beryl R. McClaskey, 19 August 1947 (MoMA: Lehmann-Haupt Papers, Box 7, Folder 2); Lutzeier, letter to Constable, July 6, 1949 (AAA: Constable Papers, 481); *HICOG Information Bulletin,* March 1950, 53.

4. Hellmut Lehmann-Haupt, "Art in Postwar Berlin," undated essay (MoMA: Lehmann-Haupt Papers, Box 1, Folder 6).

5. McClaskey, letter to Smithsonian Institution, August 16, 1947 (NGA: Cultural Relations, RG 7, Box 10).

6. Lutzeier, letter to Constable, 6 July 1949 (AAA: Constable Papers, MF #481).

7. The artists in Prolog included Ima Breusing, Josef Hegenbarth, Karl Hartung, Bernard Heiliger, Karl Hofer, Hans Jaenisch, Max Kaus, Georg Kolbe, Max Leube, Graf Luckner, Gerhard Marcks, Bruno Merbitz, Margarethe Moll, Hans Orlowski, Paul Rosie, Cornelis Ruhtenberg, Gustav Seitz, Karl Schmidt-Rottluff, Renée Sintenis, Friedrich Stabenau, Paul Strecker, Friedrich Winkler, and Mac Zimmermann, many of whom had important careers in Germany both before and after World War II. McClaskey, letter to Smithsonian Institution, 16 August 1947 (NGA: Cultural Relations, RG 7, Box 10).

8. Hellmut Lehmann-Haupt, "A Letter from America," in *Prolog: A Gift in Friendship for Beryl Rogers McClaskey and Charles Baldwin* (Berlin: Brothers Hartmann, 1948).

9. Friedrich Winkler, "Letter from Berlin," in *Prolog: A Gift in Friendship*.

10. Kurt Hartmann, "A Cup of Coffee and a Nursery Song," in *Prolog: A Gift in Friendship*.

11. Peter F. Szluk, "'Freedom' Prizes for Artists," *HICOG Information Bulletin*, June 1950.

12. Ibid.

13. The Haus am Waldsee, a small mansion built in 1922 for a wealthy Jewish family, was later the home of Karl Melzer, the president of the Reichsfilmkammer. The Berlin Magistrate transformed the house into a cultural center in January 1946. Its first director was Robert Büchner, succeeded in 1946 by the poet and art historian Dr. Karl Ludwig Skutsch. Under Skutsch, the Haus am Waldsee hosted art exhibits, poetry readings, art historical seminars, and musical events. The first exhibit showed the work of Käthe Kollwitz and Ewald Vetter (January—February 1946); subsequent exhibits featured both well-known modern artists and local artists from the Zehlendorf area. The Koerner exhibit was the tenth show of the gallery. The organizers, in addition to Lehmann-Haupt and Howard, were Sam Rosenberg and Lincoln Kirstein, both with MFA&A, and Rosemarie Müller, Lehmann-Haupt's future wife, from the Department of Art of the Berlin Magistrate. Koerner's work was not shown in Austria until 1996. See Bernd Rottenburg, *Das Haus am Waldsee unter der Leitung von L. K. Skutsch in den Jahren 1946 bis 1958* (Berlin, Diplomarbeit, Fachhochschule für Technik und Wirtschaft, 1997), and the files on postwar exhibition activities at the Haus am Waldsee Archiv.

14. Koerner never knew exactly when and where his parents and brother died. It is now known that his mother and father were deported from Vienna on June 9, 1942, and eventually taken to Mali-Trostinec, an extermination camp in Belorussia. They were gassed on June 15, 1942. Koerner's brother and sister-in-law were taken to the Kielce ghetto in Poland on February 19, 1942, and most probably murdered by the summer of 1942 in Belzec, Sobibor, or Treblinka. Most of Koerner's other relatives also perished in extermination camps in the East. Joseph Leo Koerner, *Henry Koerner 1915–1991: Unheimliche Heimat* (Vienna: Österreichische Galerie Belvedere, 1997), 21.

15. Frederick R. Brandt, "Henry Koerner," MA thesis, Pennsylvania State University, 1963, 8; Henry Koerner, quoted in Gail Stavitsky, "Of Two Worlds: The Painting of Henry Koerner," *Arts Magazine* 60 (May 1986).

16. Henry Koerner, quoted in Gail Stavitsky, *Henry Koerner* (Pittsburgh: University of Pittsburgh Press, 1983), 15.

17. "Henry Koerner: His Own Tragedy Spurs Artists to Paint Moving Postwar Pictures," *Life,* May 10, 1948; "Critics in Berlin Laud Koerner Art: American's One-Man Show Is Highly Praised by Germans," *New York Times,* April 4, 1947, 21; *Time* magazine, quoted in Stavitsky, *Henry Koerner,* 14.

18. Helen Boswell, "Berlin Newsletter," *Art Digest,* June 1, 1947; Jo Gibbs, *Art Digest,* February 1, 1948.

19. "Henry Koerner," *Heute,* May 15, 1947; Edwin Redslob, "Übersteigerte Wirklichkeit: Zur Ausstellung des amerikanischen Malers Henry Koerner," *Tagespiegel,* April 4, 1947.

20. Koerner questionnaire (MoMA: Lehmann-Haupt Papers, Box 1, Folder 3).

21. See, for instance, Charlotte Weidler, "Art in Western Germany Today," *Magazine of Art,* 1951, and Helen Boswell, "Berlin Newsletter," *Art Digest,* February 15, 1947.

22. Jost Hermand estimates that 4 to 5 percent of Western German artworks in the immediate postwar period were antifascist, 5 to 8 percent nonrepresentational, and 85 to 90 percent "half modern" or classical modern. Jost Hermand, "Freiheit im Kalten Krieg," in *'45 und die Folgen: Kunstgeschichte eines Wiederbeginns* (Koln: Bohlau Verlag, 1991). A 1946 exhibit of contemporary French prints, organized by the French military government, included thirty-two works by Léon Delarbre, a French inmate in Auschwitz, Dora, and Buchenwald. This, however, was an exception, as confirmed by the author's examination of the exhibition catalogs of the period. In *Kunst und Kunstpolitik 1945–1949,* Jutta Held insists that artists were reflecting on the war through symbolic means, but assessing this claim involves reading the artist's intentionality into the image. See also Martin Damus, *Kunst in der BRD 1945–1990* (Hamburg: Rowohlt, 1995); Marion Deshmukh, "The Year Zero and Beyond," in George O. Kent, ed., *Historians and Archivists: Essays in Modern German History and Archival Policy* (Fairfax, VA: George Mason University Press, 1991); Yule F. Heibel, *Reconstructing the Subject: Modernist Painting in Western Germany, 1945–1950* (Princeton: Princeton University Press, 1995); Jutta Held, *Kunst und Kunstpolitik 1945–1949* (Berlin: Elefanten Press, 1981); Andre Kormendi, "What German Art Survives," *Art News,* October 1948; Markus Krause, *Galerie Gerd Rosen: Die Avantgarde in Berlin 1945–1949* (Berlin: Ars Nicolai, 1995); Edouard Roditi, "The Destruction of the Berlin Museums," *Magazine of Art* 42, no. 8 (December 1949); and Eckart Gillen and Diether Schmidt, eds., *Zone 5: Kunst in der Viersektorenstadt 1945–1951* (Berlin: Dirk Nishen, 1989).

23. Artists such as Gerhard Altenbourg, Johannes Backes, Volker Bohringen, Heinrich Ehmsen, Willi Geiger, Hans Grundig, Lea Grundig, Georg Netzband, Max Radler, Herbert Sandberg, Rudolf Schlichter, Friedrich Schroder-Sonnenstern, Tisa

von Schulenberg, Karl Staudinger, Horst Strempel, and Magnus Zeller were dealing with political and social issues. See Sybil Milton and Janet Blatter, *Art of the Holocaust* (New York: Routledge, 1981).

24. Hans. J. Gaedick letter to Tom Hutton, August 5, 1948 (in English) (Landesarchiv Berlin: OMGBS, RG 260, 4/12–1/18).

25. Ibid.

26. Tom Hutton, memo, "Political Project, Graphic Artists Berlin," to Director, OMGUS Berlin, August 16, 1948 (Landesarchiv Berlin: OMGBS, RG 260, 4/12–1/8).

27. Ibid.

28. Ibid.

29. Ibid.

30. Theodore A. Heinrich, letter to Consulado General de Chile, February 11, 1949 (NARA: RG 260, Box 84, File 1).

31. Charlotte Weidler, "Art in Western Germany Today," *Magazine of Art* 44, no. 4 (April 1955): 135.

32. Joan Luckach, *Hilla Rebay: In Search of the Spirit in Art* (New York: George Braziller, 1983).

33. Heibel, *Reconstructing the Subject,* 7; Luckach, *Hilla Rebay,* 280–83. Franz Roh was a central figure in the Weimar "New Vision" movement in photography, called by his friends Willi Baumeister, George Grosz, László Moholy-Nagy and Kurt Schwitters the "Nero (Ne-Roh) of Criticism." In 1925 Roh published *Post-Expressionism–Magic Realism: Problems of Recent European Painting,* in which he coined the term "magic realism." His work was rejected by the Nazis, and in 1933 Roh was forbidden to write by government censors. In the postwar years he was the director of the Munich Cultural League, the editor of its magazine and two other art journals (*Halbjahrbuch für Alte und Neue Kunst* and *Kunst: Malerei, Plastik, Graphik, Architektur, Wohnkultur*), and a prolific writer whose articles appeared regularly in the *Neue Zeitung.* He also hosted a radio program on art on the Bayerischen Rundfunk, and and in 1951 became president of the International Association of Art Critics. "Franz Roh—Photography & Collage from the 1930s," exh. cat., Ubu Gallery, New York, September 2006.

34. Luckach, *Hilla Rebay,* 284; F. Roh and J. A. Twhaites, *"Zen 49: Erste Ausstellung im April 1950,"* exh. cat., (Munich: Amerikahaus, 1950). The Amerikahäuser, also known as U.S. Information Centers, were cultural centers established by the American military government in the main cities of the American zone and in Berlin. By 1951 there were twenty-seven Amerikahäuser in West Germany, each with an open-shelf library, English teaching facilities, concerts, lecture programs, and children's events. See "U.S. Information Centers," *HICOG Information Bulletin,* January 1950, 10–16.

35. Luckach, *Hilla Rebay,* 285, 279–80. Hellmut Lehmann-Haupt, *Art under a Dictatorship* (New York: Oxford University Press, 1954), 199.

36. Thomas Grochowiak, "Neuanfänge '45 aus der sicht der Künstlers," in

'45 und die Folgen: Kunstgeschichte eines Wiederbeginns (Köln: Bohlau Verlag, 1991), 181.

37. In 1940 Davis founded Ballet Theater, the first American ballet company, which staged more than forty ballets in ten years and employed a significant number of new American choreographers and musicians. During World War II, he directed steel and industrial pageants, and in 1942 he produced the war show *Here's Your Army,* which raised three million dollars for the army relief fund. After World War II, Davis became a board member of the American National Theater and Academy, an organization chartered by Congress, and collaborated in international cultural programs in the belief that the U.S. government should take a dominant role in disseminating American cultural advances abroad. In 1949 Davis financed a European tour of African American performers from Howard University out of his own pocket, with the goal of neutralizing the negative image of America in Europe. In 1952 he coproduced a musical based on George Gershwin's *Porgy and Bess.* This production, cosponsored by the State Department, began an extraordinarily successful international tour in 1952 and reached its peak of political impact in December 1955, when it was presented in Leningrad—the first time an American theater group was admitted into the Soviet Union. See David Monod, "He Is a Cripple an' Needs My Love: Porgy and Bess as Cold War Propaganda," in Scott-Smith and Krabbendam, *Cultural Cold War.*

38. Thomas Grochowiak, "Neuanfänger '45 aus der sicht der Künstlers," in Klaus Honnef and Hans M Schmidt, eds., *Kunst und Kultur im Rheinland und in Westfalen 1945–1952: Aus den Trümmern—Neubeginn und Kontinuität* (Bonn: Rheinland Verlag, 1985); Ursula Bluhm, memorandum, June 1, 1950 (AAA: Constable Papers, 610–12).

39. The term "classical modern" is something of an oxymoron. Modern art is by definition outside the framework of the conventional art of a given period. Classicism, on the other hand, is officially sanctioned and thus implies the existence of an authority that establishes canons of taste and procedure.

40. Franz Roh, *German Art in the 20th Century* (New York: New York Graphic Society, 1968), 252.

41. *HICOG Information Bulletin,* April 1950, inside cover.

42. Hans Belting, *The Germans and Their Art: A Troublesome Relationship* (New Haven: Yale University Press, 1998), 86.

43. "Culture and Politics," *Information Control Review* (Headquarters OMGUS Information Control Division), no. 33 (1946), 5–6.

44. The text of JCS 1779 can be found in U.S. Department of State, *Germany 1945–1949,* 33–42.

45. OMGBS Education and Cultural Relations Branch, "Long-Range Policy Statement for German Re-education," July 1947 (Landesarchiv Berlin: Box 14-3, Folder 3).

46. "Kultur and Politics," restricted Research Branch report, July 25, 1947 (Landesarchiv Berlin: OMGBS, RG 260, 4/18–3/1). The document also contains what seems

to have been OMGUS's working definition of *Kultur:* "The German word 'Kultur' (for which there is no exact equivalent in the English language) is a dangerously indefinite word. One moment it is used to signify the sum-total of civilization, including its moral standards (as when the political parties promise to protect Christian-Western 'Kultur'); the next moment it means no more than an education (as when the word 'Kultur' appears on a poster advertising an illustrated lecture on Bulgaria).... This diversity of meanings permits many groups to consider themselves the guardians of 'Kultur.' Because of the noble connotations of the word, even the simplest activity, when described as 'Kultur' becomes strangely sanctified in the popular mind. Today many hope to save 'Kultur', and many hope to be saved by it. For some, it is the last refuge in defeat and disaster."

47. Constable, memorandum, *Art and Reorientation,* 1949 (AAA: Constable Papers, MF #781–90); Public Information Office, OMGUS Bad Nauheim, Report, 10 June 1949 (AAA: Constable Papers, MF #458). See also Constable, memorandum to Dr Alonzo Grace, 1 April 1949 (AAA: Constable Papers, MF #364–65). For more on Constable's activities in Germany, see Cora Sol Goldstein, "Power and the Visual Domain: Images, Iconoclasm, and Indoctrination in American Occupied Germany, 1945–1949," PhD diss., University of Chicago, 2002; Heibel, *Reconstructing the Subject;* and Sigrid Ruby, *Have We an American Art?* (Weimar: Verlag und Datenbank für Geisteswissenschaften, 1999).

48. Constable, memorandum, *Art and Reorientation,* 1949 (AAA: Constable Papers, 781–90).

Chapter Five

1. See Dario Gamboni, *The Destruction of Art: Iconoclasm and Vandalism since the French Revolution* (New Haven: Yale University Press, 1997).

2. Paul M. A. Linebarger, *Psychological Warfare* (Washington, DC: Combat Forces Press, 1948), 67–68.

3. Morris L. Ernst and Pare Lorentz, *Censored: The Private Life of the Movie* (New York: Jonathan Cape and Harrison Smith, 1930); Thomas Doherty, *Projections of War: Hollywood, American Culture, and World War II* (New York: Columbia University Press, 1993).

4. Elmer Davis and Byron Price, *War Information and Censorship* (Washington, DC: American Council on Public Affairs, 1943); Clayton D. Laurie, *The Propaganda Warriors: America's Crusade against Nazi Germany* (Lawrence: University Press of Kansas, 1996); Allan Winkler, *The Politics of Propaganda: The Office of War Information, 1942–1945* (New Haven: Yale University Press, 1978); Michael S. Sweeney, *Secrets of Victory: The Office of Censorship and the American Press and Radio in World War II* (Chapel Hill: University of North Carolina Press, 2001).

5. The text of the Yalta communiqué can be found in Charles I. Bevans, ed. *Treaties and Other International Agreements of the United States of America* (Washington, DC: U.S. Government Printing Office, 1969), 3:10005.

6. The text of Control Council Directive No. 30 can be found in John Paul Weber, *The German War Artists* (South Carolina: Cerberus, 1979). For its implementation by OMGUS, see "Demilitarization (Cumulative Review)," *Monthly Report of the Military Governor, U.S. Zone,* no. 13, August 20, 1946.

7. Many buildings constructed in the Nazi period had been destroyed by Allied bombing, but those that remained in reasonable condition were repaired and "normalized"— cleansed of Nazi emblems and given a new function. The gigantic eagles adorning the façade of leading Nazi art museum, the Haus der Deutschen Kunst in Munich, were replaced with the American flag, and the building was transformed into an officers club. The Munich *Führerbau,* the central administrative building of the Nazi Party, became first a depository for looted art and later the Munich America House. Tempelhof airport in Berlin, the center of operations for the Luftwaffe (and the largest airport in the world at the time), continued to operate under American control. The large black eagle that adorned the main entrance remained in place, though in modified form—the white swastika on its breast was replaced by an American striped shield, and the bird's head was painted white. See Gavriel D. Rosenfeld, *Munich and Memory: Architecture, Monuments, and the Legacy of the Third Reich* (Berkeley: University of California Press, 2000), and Thomas Parrish, *Berlin in the Balance, 1945–1949: The Blockade, the Airlift, the First Major Battle of the Cold War* (Reading. Massachusetts: Perseus Books, 1998), 236.

8. JCS 1067, U.S. Department of State, *Germany 1945–1949: The Story in Documents* (Washington, DC: U.S. Government Printing Office, 1950).

9. Law No. 52, Military Government—Germany, U.S. Zone, July 14, 1945. Text in Weber, *German War Artists,* 19–22.

10. The text of Title 18 can be found in Weber, *German War Artists,* 23–27.

11. Ibid., 36; Gordon W. Gilkey, "German War Art," Office of the Chief Historian Headquarters, European Command, April 25, 1947; Memorandum "Returned of Captured Paintings to Germany," from E. M. Harris, Lt. Colonel, G.S.C., executive, to the Judge Advocate General of the U.S. Army, November 6, 1950 (NARA: RG 319. Box 72, Folder 17).

12. Harry H. Almond Jr., Office of Assistant General Counsel, International Affairs, "Memorandum from the Department of Defense for the General Counsel on Disposal of German War Art," February 27, 1978. U.S. Army Center of Military History Archives. See also Marion F. Deshmukh, "The German War Art Collection," in Geoffrey J. Giles, ed., *Archivists and Historians* (Washington, DC: German Historical Institute, 1996). Gilkey organized a small exhibit (103 pieces) of German war art for the U.S. Army personnel at the Staedel Museum in Frankfurt am Main, before shipping the collection out of Germany. Galen L. Geer, "Nazi War Art," *Gung-Ho: The Magazine of the International Military Man,* October 1981, 34–59; Gilkey, "German War Art."

13. Gilkey, "German War Art," 27. Project No. NND775057, Property Cards: Nazi and Militaristic Property Transfers 1945–1951 (NARA: RG 260, Box 232). Some of the Nazi paintings found their way into the offices of high military officers in Wash-

ington (NARA: RG 319, Box 72, Folder 17). Memo from Lt. Colonel Harold R. James, CSC, Acting Executive, Special Staff, U.S. Army, Washington, D.C., to Major General Orlando Ward, Chief, Military History, Special Staff, U.S. Army, 21 September 1950. Two Nazi originals hung in General Ward's office, and one in Lt. Colonel James's (NARA: RG 319, Box 72, Folder 14).

14. Title 18, in Weber, *German War Artists*, 23. The American personnel who determined what to preserved often knew little about Nazism. Their most blatant mistake was the treatment of German folklore. By the end of June 1946, when military museums and public art collections were still closed, the Allied Control Authority authorized the reopening of "provincial museums of local interest . . . and museums dedicated to the memory of the progressive benefactors of the German race" (Historical Report of the Operation of the Office of Military Government Berlin District, July 1, 1945–June 30, 1946, Vol XI, Monuments, Fine Arts and Archives [Landesarchiv Berlin: RG 260, B8-2, Folder 1]). The reference to the "German race" is striking, given that the Allies specifically forbade the exhibition of objects or documents promoting "racially prejudiced ideology." The American iconoclasts apparently did not recognize that folklore had been a cornerstone of Nazi ideology.

15. Rebecca Boehling, *A Question of Priorities: Democratic Reforms and Economic Recovery in Postwar Germany* (New York: Berghahn Books, 1996), 26; Schivelbusch, *In a Cold Crater,* 33.

16. Edward C. Breitenkamp, *The U.S. Information Control Division and Its Effect on German Publishers and Writers 1945 to 1949* (Grand Forks, ND: University Station, 1953), 40–41.

17. Report, Public Information Office, OMGUS, 16 December 1948 (Landesarchiv Berlin: OMGUS [Berlin], RG 260, MF # 4/1–3/9).

18. By January 1947 ICD had distributed eighty-nine press licenses. Licensees included thirty-eight SPD members, twenty-four CDU/CSU supporters, and four KPD supporters. The breakdown by religion was thirty-three Catholic, twenty-eight Protestant, three Jewish, one Unitarian, and twenty-four nonreligious. Larry Hartenian, *Controlling Information in U.S. Occupied Germany, 1945–1949: Media Manipulation and Propaganda* (Lewiston: Edwin Mellen Press, 2003), 115–16, 127. See also Norbert Frei, *Amerikanische Lizenzpolitik und deutsche Pressetradition: Die Geschichte der Nachkriegszeitung Suedost-Kurier* (Munich: Oldenbourg Verlag, 1986).

19. The *Ulenspiegel* is named after Till Eulenspiegel, a joker and jester in German folklore. This astute peasant was a populist hero who played practical jokes throughout Europe, respecting neither nobility nor the pope. The literal translation of the name is "owl mirror," but the Low German version, *ul'n Spegel,* means "wipe the arse." The adventures of Till Eulenspiegel were translated into Dutch, French, English, Latin, Danish, Swedish, Bohemian and Polish. The *Ulenspiegel* was not the only satirical journal in occupied Germany: OMGUS also licensed *Wespennest: Politischsatirischen Wochenzeitschrift* in Stuttgart and *Simpl* in Munich in 1946. The British licensed *Bumerang: Das Blatt, das keiner ernst nimmt* and *Puck,* the French licensed

Michel, and SMAD launched *Frische Wind* in 1947. See Bernhard Jendricke, *Die Nachkriegszeit im Spiegel der Satire: Die satirischen Zeitschriften Simpl und Wespennest in den Jahren 1946 bis 1950* (Frankfurt am Main: Peter Lang, 1982).

20. Herbert Sandberg, *Spiegel Eines Lebens* (East Berlin: Aufbau Verlag, 1988), 62. See also Herbert Sandberg, *Ulenspiegel: Deutschland vor der Teilung* (Oberhausen: Ludwig Institut Schloß Oberhausen, 1990); Herbert Sandberg, *Herbert Sandberg: Leben und Werk* (East Berlin: Henschelverlag, 1977); Herbert Sandberg, *Herbert Sandberg: 40 Jahre Graphik und Satire, Ausstellung zum 60 Geburstag* (East Berlin: Staatlichen Museum Berlin, 1968); and Herbert Sandberg and Gunter Kunert, eds.,*Ulenspiegel: Zeitschrift fur Lieratur, Kunst und Satire1945-1950* (Berlin: Ulenspiegel Verlag, 1988).

21. Ann Taylor Allen, *Satire and Society in Wilhelmine Germany: Kladderadatsch and Simplicissimus* (Lexington: University Press of Kentucky, 1984), 37, 43; author's interview with Lilo Sandberg, wife of Herbert Sandberg, Berlin, 1999.

22. Emil Carlebach (1914–2001) was born into a middle-class Jewish family from Frankfurt. In 1932 he became a member of the Kommunistische Partei Deutschlands (KPD). In 1934 he was imprisoned by the Nazis and was then sent to Dachau in 1937 and transferred to Buchenwald in 1938. After liberation Carlebach returned to Frankfurt, where he served in the city council and participated in the reconstruction of the KPD. In 1956, when the KPD was banned in West-Germany, he fled to East Germany, where he worked as a journalist. In 1969, as a Social Democratic government in West Germany allowed the foundation of a new Communist Party, the Deutsche Kommunistische Partei, Carlebach returned to West Germany. See Emil Carlebach, *Zensur ohne Schere: Die Gründerjahre der Frankfurter Rundschau 1945/47* (Frankfurt am Main: Röderberg Verlag, 1985).

23. Sandberg, *Ulenspiegel,* 140. Johannes R. Becher (1891–1958) was a German Communist agitator and propagandist who had been exiled in Moscow since 1934. In July 1945 Becher founded the Kulturbund für die demokratische Erneuerung Deutschlands (Cultural Union for the Democratic Renewal of Germany). The Kulturbund had been planned by the Ulbricht group in Moscow. It was to follow the Communist model—an institution under a supposedly democratic, nonpartisan administration but controlled politically by the Communists. In November 1947 OMGUS closed the Kulturbund in the American zone and sector. In 1954, Becher became the minister of culture of the GDR. Bernard Genton, *Les alliés et la culture: Berlin, 1945–1949* (Paris: Presses Universitaires de France, 1998), 32–51.

24. Sandberg, *Spiegel Eines Lebens,* 65. On de Mendelssohn, see Wolfgang Schivelbusch, *In a Cold Crater: Cultural and Intellectual Life in Berlin,1945–1948* (Berkeley: University of California Press, 1998), 155–61.

25. Genton, *Les alliés et la culture,* 279–312. See also Wulf Koepke, "Günther Weisenborn's Ballad of his Life," in Neil H. Donahue and Doris Kirchner, eds., *Flight of Fantasy: New Perspectives on Inner Emigration in German Literature, 1933–1945* (New York: Berghahn Books, 2003); Günther Weisenborn, *Der Gespaltene Horizont* (Munich: Desch, 1964).

26. The Russian Bolshevik leadership in 1917 was split between those who had spent most of their lives in Russian underground organizations, in tsarist prisons, and in the Siberian exile, and those who had lived in exile, in Europe or the United States. The future "nativists" (Stalin, Dzerzhinsky) acquired a particular mentality—paranoiac, party-centered, xenophobic, and certainly isolationist—and held a grudge against the future "cosmopolitans" (Trotsky, Bukharin, Kollantai). A similar division can be seen in the 1945–1955 period, when the German Communist leadership was split between veterans of the Moscow exile, who remained loyal to the Kremlin (e.g., Ulbricht) and those who had gone to Mexico, knew the United States, and were purged from the party and German political life in the 1950s. See Jeffrey Herf, *Divided Memory: The Nazi Past in the Two Germanys* (Cambridge: Harvard University Press, 1997); and Orlando Figes, *A People's Tragedy: The Russian Revolution 1891–1924* (New York, 1996), 296–97.

27. Although concentration and extermination camps figured prominently in the *Ulenspiegel* iconography of 1946, covers and full-page images made no reference to the racialist character of the Nazi regime or to its war against European Jews.

28. *Ulenspiegel,* year 1, no. 1, December 1945 ("Architecture"); year 1, no. 6, March 1946 ("Assault"); year 1, no. 22, November 1946 ("Moloch").

29. *Ulenspiegel,* year 1, no. 2, January 1946 ("Anti-fascists"); year 1, no. 13, June 1946 ("Fragebogen").

30. *Ulenspiegel,* year 1, no. 3, January 1946 ("Three Holy Kings"); year 1, no. 17, August 1946 ("German Buildings"); *Ulenspiegel,* year 1, no. 24, November 1946 ("He lives under a better sky").

31. *Ulenspiegel,* year 1, no. 1, December 1945 ("Christmas"); year 1, no. 8, April 1946 ("Tiergarten"); year 1, no. 7, March 1946 ("Guilt").

32. See Jessica C. E. Gienow-Hecht, *Transmission Impossible: American Journalism and Cultural Diplomacy in Postwar Germany, 1945–1955* (Baton Rouge: Louisiana State University Press, 1999). Also see Hartenian, *Controlling Information,* 152–54; and Jessica C. E. Gienow-Hecht, "Friends, Foes, or Reeducators? Feinbilder and Anticommunism in the U.S. Military Government in Germany, 1946–1953," in Ragnhild Fiebig-von Hase and Ursula Lehmkuhl, eds., *Enemy Images in History* (Providence: Berghahn Books, 1997), 281–301.

33. Hartenian, *Controlling Information,* 179. See also Cedric Belfrage, *Seeds of Destruction* (New York: Cameron and Kahn, 1954).

34. Wallenberg left the *Neue Zeitung* in August 1947 and was replaced by Jack Fleischer, a former correspondent for Henry Luce's anticommunist publications, *Life* and *Time.* Fleischer intensified the anti-Soviet perspective of the newspaper. By mid-1948, however, ICD reproached Fleischer for the weakness of his coverage of American news and replaced him with Kendall Foss, a fervent anticommunist. Foss, however, was immediately criticized for allowing anti-American ideas in the *Neue Zeitung.* Finally, in February 1949, three rabidly anticommunist Americans, John Elliot, Marcel W. Fodor (born in Hungary), and Jack M. Stewart, were placed over

Foss to ensure that the newspaper became a real mouthpiece for OMGUS in preparation for the imminent elections.

35. See Clare Flanagan, *A Study of German Political-Cultural Periodicals from the Years of Allied Occupation, 1945–1949* (Lewiston: Edwin Mellen Press, 2000), 151–82.

36. Hartenian, *Controlling Information,* 179–80.

37. *Ulenspiegel,* year 2, no. 14, July 1947 ("Wiedergutmachung"); year 2, no. 25, December 1947 ("With Other Eyes").

38. *Ulenspiegel,* year 2, no. 8, April 1947.

39. *Ulenspiegel,* year 2, no. 7, April 1947 ("Conversation"); year 2, no. 18, September 1947 ("Fifty-Fifty"); year 2, no. 20, October 1947 ("Right Way").

40. Hartenian, *Controlling Information,* 222.

41. François Furet, *Le passé d'une illusion. Essai sur l'ideé communiste au XX siècle* (Paris: Robert Laffont, 1995), 645.

42. Information Control, "Monthly Report," May 3, 1948 (Landesarchiv Berlin: OMGBS, RG 260, 5/39–1/18).

43. Ibid.

44. Hartenian, *Controlling Information,* 301.

45. See Brian T. van Sweringen, *Kabarettist an der Front des Kalten Krieges: Günter Neumann und das politische Kabarett in der Programmgestaltung des Radios im amerikanischen Sektor Berlins (RIAS)* (Passau: Richard Rothe, 1989), 50, 130–32; Regina Stürickow, *Der Insulaner verliert Der Ruhe nicht: Günter Neumann und seine Kabarett zwischen Kaltem Krieg und Wirtschaftwunder* (Berlin: Arani, 1993), 23–29.

46. Frances Stonor Saunders, *The Cultural Cold War: The CIA and the World of Arts and Letters* (New York: New Press, 1999), 7–32. Also see Giles Scott-Smith, "A Radical Democratic Political Offensive: Melvin J. Lasky, Der Monat, and the Congress for Cultural Freedom," *Journal of Contemporary History* 35, no. 2 (April 2000): 263–80.

47. "Monthly Report of the Military Government," September 1948, no. 39 (Landesarchiv Berlin: OMGBS, RG 260, 4/1–3/9).

48. Furet, *Le passé d'une illusion,* 642.

49. See Frederick C. Barghoorn, *Soviet Foreign Propaganda* (Princeton: Princeton University Press, 1964); Frederick C. Barghoorn, *The Soviet Image of the United States: A Study in Distortion* (New York: Harcourt Brace, 1950); W. Phillips Davison, "An Analysis of the Soviet-Controlled Berlin Press," *The Public Opinion Quarterly,* 11(1), 1947, 40–57; Alexander Dallin, "America Through Soviet Eyes," *The Public Opinion Quarterly,* 11(1), 1947, 26–39.

50. See S. Jonathan Wiesen, *West German Industry and the Challenge of the Nazi Past, 1945–1955* (Chapel Hill: University of North Carolina Press, 2001); Herf, *Divided Memory,* 201–66; and Norbert Frei, *Adenauer's Germany and the Nazi Past: The Politics of Amnesty and Integration* (New York: Columbia University Press, 2002).

51. *Tägliche Rundschau,* October 9, 1948 ("Renazification in West Germany");

December 4, 1948 ("When dogs are allowed to bark, they think they are free"); and December 7, 1948 ("Clay").

52. *Ulenspiegel,* year 3, no. 24, December 1948.

53. *Ulenspiegel,* year 4, no. 23, December 1949.

54. *Ulenspiegel,* year 4, no. 6, August 1949 ("This is Democracy"); year 4, no. 21, November 1949 ("Bonn").

55. Laszlo Rajk, the former interior minister of Hungary, a veteran of the Spanish Civil War, and the leader of the anti-Nazi resistance in Hungary, was tried in Budapest in September 1949. After a one-week show trial Rajk and four others defendants were found guilty and sentenced to death.

56. David Pike, *The Politics of Culture in Soviet-Occupied Germany, 1945–1949* (Stanford: Stanford University Press, 1992), 613.

57. Ibid., 629.

58. Sandberg, *Spiegel Eines Lebens,* 79.

59. Sandberg, *Ulenspiegel,* 148; Sandberg, *Spiegel Eines Lebens,* 79, 97, 99.

Conclusion

1. Gerda Breuer, ed., *Die Zähmung der Avantgarde: Zur Rezeption der Moderne in den 50er Jahren* (Basel: Stroemfeld Verlag, 1997), 151; Everett C. Hughes "German Diary;" 1948 (University of Chicago Archives: Everett C. Hughes Papers, Box 94).

2. Jeffrey Herf, *Divided Memory: The Nazi Past in the Two Germanies* (Cambridge: Harvard University Press, 1997), 203.

3. See Peter Novick, *The Holocaust in American Life* (Boston: Houghton Mifflin, 1999), 213; Siegfried Zielinski and Gloria Custance, "History as Entertainment and Provocation," *New German Critique* 19, no. 1 (Winter 1980): 81–96; and Mark E. Cory, "Some Reflections on NBC's Film Holocaust," *German Quarterly* 53, no. 4 (November 1980): 444–51.

4. Konrad H. Jarausch, *After Hitler: Recivilizing Germans, 1945–1955* (Oxford: Oxford University Press, 2006), 66–69.

Bibliography

Abzug, Robert H., ed. *GIs Remember: Liberating the Concentration Camps.* Washington, DC: National Museum of American Jewish Military History, 1994.

———. *Inside the Vicious Heart: Americans and the Liberation of Nazi Concentration Camps.* Oxford: Oxford University Press, 1985.

Adam, Peter. *Art of the Third Reich.* New York: Harry N. Abrams, 1992.

Alexander, Edward. *The Resonance of Dust: Essays on Holocaust Literature and Jewish Fate.* Columbus: Ohio State University Press, 1979.

Alexander, William. *Film on the Left: American Documentary Film from 1931 to 1942.* Princeton: Princeton University Press, 1981.

Allen, Ann Taylor. *Satire and Society in Wilhelmine Germany: Kladderadatsch and Simplicissimus.* Lexington: University Press of Kentucky, 1984.

Allen, Richard. *Projecting Illusion: Film Spectatorship and the Impression of Reality.* Cambridge: Cambridge University Press, 1995.

Ambrose, Stephen E. *Citizen Soldiers: The U.S. Army from the Normandy Beaches to the Bulge to the Surrender of Germany.* New York : Simon & Schuster, 1997.

Amtliches Material zum Massenmord von Katyn. Berlin: Gedruckt im Deutschen Verlag, 1943.

Anfam, David. *Abstract Expressionism.* New York: Thames and Hudson, 1994.

Annan, Noel. *Changing Enemies: The Defeat and Regeneration of Germany.* Ithaca: New York, 1995.

Assmann, Aleida. "On the (In)Compatibility of Guilt and Suffering in German Memory." *German Life and Letters* 59, no. 2 (April 2006): 187–200.

Atrocities: A Study of German Reactions. Supreme Headquarters Allied Expeditionary Force, Psychological Warfare Division, Intelligence Section, June 21, 1945.

Atrocities and Other Conditions in Concentration Camps in Germany. Report of the Committee Requested by General Dwight D. Eisenhower through the Chief of Staff, General George C. Marshall, to the Congress of the United States, Presented by Mr. Barkley, May 15, 1945. Washington, DC: U.S. Government Printing Office.

Babelon, Jean-Pierre, and André Chastel. *La notion de patrimoine.* Paris: Liana Levi, 1996.

Bacevich, Andrew J. *American Empire: the Realities and Consequence of U.S. Diplomacy.* Cambridge: Harvard University Press, 2002.

Bance, Alan, ed. *The Cultural Legacy of the British Occupation in Germany*. Stuttgart: Hans-Dieter Heinz, 1997.

Bargohoorn, Frederick C. *Soviet Foreign Propaganda*. Princeton: Princeton University Press, 1964.

———. *The Soviet Image of the United States: A Study in Distortion*. New York: Harcourt Brace, 1950.

Barnouw, Dagmar. *Germany 1945: Views of War and Violence*. Bloomington: Indiana University Press, 1996.

Barnouw, Erik. *Documentary: A History of the Non-fiction Film*. New York: Oxford University Press, 1993.

Barkan, Elazar. *The Retreat of Scientific Racism: Changing Concepts of Race in Britain and the United States between the World Wars*. New York: Cambridge University Press, 1992.

Barr, Alfred H., Jr. "Art in the Third Reich—Preview 1933." *Magazine of Art* 38, no. 6 (October 1945): 212–22.

Barsam, Richard Meran, ed. *Nonfiction Film: Theory and Criticism*. New York: E. P. Dutton, 1976.

Bausch, Ulrich M. *Die Kulturpolitik der US-amerikanischen Information Control Division in Württemberg-Baden von 1945 bis 1949*. Stuttgart: Klett-Cotta, 1992.

Belfrage, Cedric. *Seeds of Destruction*. New York: Cameron and Kahn, 1954.

Belting, Hans. *The Germans and Their Art: A Troublesome Relationship*. New Haven: Yale University Press, 1998.

Bendersky, Joseph W. *The "Jewish Threat": Anti-Semitic Politics of the U.S. Army*. New York: Basic Books, 2000.

———. "From Cowards and Subversives to Aggressors and Questionable Allies: U.S. Army Perceptions of Zionism since World War I." *Journal of Israeli History* 25, no. 1 (2006): 107–29.

———. "The Absent Presence: Enduring Images of Jews in United States Military History." *American Jewish History* 89, no. 4 (2001): 411–36.

Benton, William. "Advancing American Art." *Art News* 45, no. 8 (October 1946): 19.

Benz, Wolfgang. *Potsdam 1945: Besatzungsherrschaft und Neuaufbau im Vier-Zonen-Deutschland*. Munich: Deutscher Taschenbuch Verlag, 1986.

Berger, John. *About Looking*. New York: Vintage, 1980.

Berghahn, Daniela. *Hollywood behind the Wall: The Cinema of East Germany*. Manchester: Manchester University Press, 2005.

Berghahn, Volker R. *America and the Intellectual Cold Wars in Europe*. Princeton: Princeton University Press, 2001.

———. "Conceptualizing the American Impact on Germany: West German Society and the Problem of Americanization." Presented at "The American Impact on Western Europe: Americanization and Westernization in Transatlantic Perspective." Conference at the German Historical Institute, Washington, DC, March 25–27, 1999.

Bevans, Charles I., ed. *Treaties and Other International Agreements of the United States of America*. Vol. 3. Washington, DC: U.S. Government Printing Office, 1969.

Biberman, Edward. "The Attack on the American Artist." Conference of the Hollywood Arts, Science, and Professions Council, July 9–13, 1947. Hollywood, 1947.

Black, Gregory D. *The Catholic Crusade against the Movies, 1940–1975.* Cambridge: Cambridge University Press, 1997.

Bloxham, Donald. *Genocide on Trial: War Crimes Trials and the Formation of Holocaust History and Memory.* Oxford: Oxford University Press, 2001.

Blum, John Morton. *V Was for Victory: Politics and American Culture during World War II.* San Diego: Harcourt Brace Jovanovich, 1976.

Blumenson, Martin. *The Patton Papers, 1940–1945.* Boston: Houghton Mifflin, 1974.

Boehling, Rebecca. *A Question of Priorities: Democratic Reforms and Economic Recovery in Postwar Germany.* New York: Berghahn Books, 1996.

———. "The Role of Culture in American Relations with Europe: The Case of the United States Occupation of Germany." *Diplomatic History* 23, no. 1 (Winter 1999): 57–69.

Borger, Hugo, Mai Ekkerhardt, and Stephan Waetzoldt, eds. *'45 und die Folgen: Kunstgeschichte eines Wiederbeginns.* Cologne: Bohlau Verlag, 1991.

Borstelmann, Thomas. *The Cold War and the Color Line: American Race Relations in the Global Arena.* Cambridge: Harvard University Press, 2001.

Boswell, Helen. "Berlin Newsletter." *Art Digest* 21, no. 17 (June 1, 1947): 17.

———. "Berlin Newsletter." *Art Digest* 22 (November 15, 1947): 4.

———. "Letter from Berlin." *Art Digest* 21 (February 15, 1947): 22.

Boswell, Peyton. "When Art Becomes 'War Surplus.'" *Art Digest* 22, no. 17 (June 1, 1948): 7.

Bourke-White, Margaret. *Dear Fatherland, Rest Quietly: A Report on the Collapse of Hitler's "Thousand Years."* New York: Simon & Schuster, 1946.

Braden, Thomas W., and Stewart Alsop. *OSS—Sub Rosa: The OSS and American Espionage.* New York: Reynal and Hitchcock, 1946.

Braisted, Paul J., ed. *Cultural Affairs and Foreign Relations.* Washington, DC: Columbia Books, 1968.

Brandt, Frederick R. "Henry Koerner." MA thesis, Pennsylvania State University, 1963.

Breitenkamp, Edward C. *The U.S. Information Control Division and Its Effect on German Publishers and Writers 1945 to 1949.* Grand Forks, ND: University Station, 1953.

Breuer, Gerda, ed. *Die Zähmung der Avantgarde: Zur Rezeption der Moderne in den 50er Jahren.* Basel: Stroemfeld Verlag, 1997.

Brink, Cornelia. *Ikonen der Vernichtung: Öffentlicher Gebrauch von Fotografien aus nationalsozialistischen Konzentrationslagern nach 1945.* Berlin: Akademie Verlag, 1998.

Burger, Hanus. *Der Fruhling war es wert: Erinnerung.* Munich: C. Bertelsman Verlag, 1977.

Burleigh, Michael. *Ethics and Extermination: Reflections on Nazi Genocide.* Cambridge: Cambridge University Press, 1997.

Burstow, Robert. "The Limits of Modernist Art as a 'Weapon of the Cold War': Reassesing the Unknown Prisoner Patron of the Monument to the Unknown Political Prisoner." *Oxford Art Journal* 20, no. 1 (1997): 68–80.

Caldwell, William J. "Democracy Comes to Dachau." *HICOG Information Bulletin,* May 1951, 23–25.

Carlebach, Emil. *Zensur ohne Schere: Die Gründerjahre der Frankfurter Rundschau 1945/47.* Frankfurt am Main: Röderberg Verlag, 1985.

Cassou, Jean. *Le pinture française moderne.* Berlin: Group Français du Conseil de Controle Division "Education et Affaires Culturelles," 1946.

Caute, David. *The Dancer Defects: The Struggle for Cultural Supremacy during the Cold War.* Oxford: Oxford University Press, 2005.

Chalou, George C., ed. *The Secrets War: The Office of Strategic Services in World War II.* Washington, DC: National Archives and Records Administration, 1992.

Chamberlain, Brewster S. *Kultur auf Trümmern: Berliner Berichte der amerikanischen Information Control Section Juli–Dezember 1945.* Stuttgart: Deutsche Verlags-Anstalt, 1979.

Chandler, Charlotte. *Nobody's Perfect, Billy Wilder.* New York: Schuster & Schuster, 2002.

Clark, Toby. *Art and Propaganda in the Twentieth Century: The Political Image in the Age of Mass Culture.* New York: Harry N. Abrams, 1997.

Clay, Lucius D. *Decision in Germany.* Melbourne: William Heinemann, 1950.

Cohen, Emanuel. "Film Is a Weapon." *Business Screen* 7, no. 1 (January 1948): 43.

Cohen, Karl F. *Forbidden Animation: Censored Cartoons and Blacklisted Animations in America.* Jefferson: McFarland & Company, 1977.

Cohen, Milton. "Fatal Symbiosis: Modernism and WWI." *War, Literature, and the Arts* 8 (1996): 1–47.

Cole, Harry L., and Albert K. Weinberg. "The Protection of Historical Monuments and Art Treasures." In *Civil Affairs: Soldiers Becoming Governors.* Special Studies Series on the United States Army in World War II. Washington, DC: Office of the Chief of Military History, Department of the Army, 1964.

Collins, Kathleen. "Portraits of Slave Children." *History of Photography: An International Quarterly* 9, no. 3 (1985): 187–210.

Confino, Alon. "Remembering the Second World War, 1945–1965: Narratives of Victimhood and Genocide." *Cultural Analysis* 4 (2005): 47–75.

Cory, Mark E. "Some Reflections on NBC's Film Holocaust." *German Quarterly* 53, no. 4 (November 1980):444–51.

Crew, David. "Remembering German Pasts: Memory in German History, 1871–1989." Review Article. *Central European History* 33, no. 2 (2000).

"Crime and Punishment." *Army Talks* 4, no. 9 (July 10, 1945).

"Critics in Berlin Laud Koerner Art: American's One-Man Show Is Highly Praised by Germans." *New York Times,* April 4, 1947.

Crowe, Cameron. *Conversations with Wilder.* New York: Alfred A. Knopf, 1999.

"Culture and Politics." *Information Control Review* (Headquarters OMGUS Information Control Division), no. 33, 1946.

Dalfiume, Richard M. *Desegregation of the U.S. Armed Forces*. Columbia: University of Missouri Press, 1969.

Dallin, Alexander. "America through Soviet Eyes." *Public Opinion Quarterly* 11, no. 1 (Spring 1947): 26–39.

Damus, Martin. *Kunst in der BRD 1945–1990*. Hamburg: Rowohlt, 1995.

Daugherty, William E., and Morris Janowitz, eds. *A Psychological Warfare Casebook*. Baltimore: Johns Hopkins University Press, 1958.

Davidson, Eugene. *The Death and Life of Germany: An Account of the American Occupation*. New York: Alfred A. Knopf, 1959.

Davidson, J. Leroy. "Advancing American Art." *Art News* 45, no. 8 (October 1946): 19.

Davidson, W. Phillips. "An Analysis of the Soviet-Controlled Berlin Press." *Public Opinion Quarterly* 11, no. 1 (Spring 1947): 40–57.

Davis, Elmer, and Byron Price. *War Information and Censorship*. Washington, DC: American Council on Public Affairs, 1943.

de Grazia, Victoria. *Irresistible Empire: America's Advance through 20th-Century Europe*. Cambridge: Harvard University Press, 2005.

Delage, Christian. *La vision nazie de l'histoire a travers le cinéma documentaire du Troisième Reich*. Laussane: Editions L'Age du Homme, 1989.

"Demilitarization (Cumulative Review)." *Monthly Report of the Military Governor, U.S. Zone*, no. 13, August 20, 1946.

Denison, Carol H. "First Things First." *Film News* 9, no. 5 (January 1949).

Dennis, James M. *Renegade Regionalists: The Modern Independence of Grant Wood, Thomas Hart Benton, and John Curry*. Madison: University of Wisconsin Press, 1998.

Deshmukh, Marion F. "Recovering Culture: the Berlin National Gallery and the U.S. Occupation 1945–1949." *Central European History* 27, no. 4 (1994).

Doherty, Thomas. *Projections of War: Hollywood, American Culture, and World War II*. New York: Columbia University Press, 1993.

Donahue, Neil H., and Doris Kirchner, eds. *Flight of Fantasy: New Perspectives on Inner Emigration in German Literature, 1933–1945*. New York: Berghahn Books, 2003.

Douglas, Lawrence. "Film as Witness: Screening *Nazi Concentration Camps* before the Nuremberg Tribunals." *Yale Law Journal* 105 (November 1995): 449–75.

———. "The Shrunken Head of Buchenwald: Icons of Atrocity at Nuremberg." *Representations* 63 (Summer 1998): 39–64.

Dudziak, Mary L. *Cold War Civil Rights: Race and the Image of American Democracy*. Princeton: Princeton University Press, 2000.

Edelman, Murray. *From Art to Politics: How Artistic Creations Shape Political Conceptions*. Chicago: University of Chicago Press, 1995.

Ellis, Jack C., and Betsy A. McLane. *A New History of Documentary Film*. New York: Continuum, 2005.

Erenberg, Lewis A., and Susan E. Hirsch, eds. *The War in American Culture: Society and Consciousness during WWII*. Chicago: University of Chicago Press, 1996.

Ermarth, Michael. *America and the Shaping of German Society, 1945–1955*. Providence: Berg, 1993.

Ernst, Morris L., and Pare Lorentz. *Censored: The Private Life of the Movie*. New York: Jonathan Cape and Harrison Smith, 1930.

Farago, Ladislas, ed. *German Psychological Warfare*. New York: Arno Press, 1942.

Fehrenbach, Heide. *Cinema in Democratizing Germany: Reconstructing National Identity after Hitler*. Chapel Hill: University of North Carolina Press, 1995.

"Films, Radio to Bombard Nazis." *Variety* 158, no. 7 (April 25, 1945): 2.

Flanagan, Clare. *A Study of German Political-Cultural Periodicals from the Years of Allied Occupation, 1945–1949*. Lewiston: Edwin Mellen Press, 2000.

Fralin, Frances. *The Indelible Image: Photographs of War—1846 to the Present*. New York: Harry N. Abrams, 1985.

Frassanito, William A. *Gettysburg: A Journey in Time*. New York: Charles Scribner's Sons, 1975.

Frederiksen, Oliver J. *The American Military Occupation of Germany, 1945–1953*. United States Army, Europe, Headquarters, Historical Division, 1953.

Freedberg, David. *The Power of Images: Studies in the History and Theory of Response*. Chicago: University of Chicago Press, 1989.

Frei, Norbert. *Adenauer's Germany and the Nazi Past: The Politics of Amnesty and Integration*. New York: Columbia University Press, 2002.

———. *Amerikanische Lizenzpolitik und deutsche Pressetradition: Die Geschichte der Nachkriegszeitung Suedost-Kurier*. Munich: Oldenbourg Verlag, 1986.

Friedlander, Saul, ed. *Probing the Limits of Representation: Nazism and the Final Solution*. Cambridge: Harvard University Press, 1992.

Friedrich, Carl J. "Military Government and Dictatorship." *Annals of the American Academy of Political and Social Science* 267 (January 1950): 1–7.

———. *Tradition and Authority*. New York: Praeger, 1972.

———. "Military Government as a Step toward Self-Rule." *Public Opinion Quarterly* 7, no. 4 (Winter 1943): 527–41.

Furet, François. *Le passé d'une illusion: Essai sur l'ideé communiste au XX siècle*. Paris: Robert Laffont, 1995.

———, ed. *Patrimoine, Temps, Espace*. Paris: Fayard, 1997.

Gamboni, Dario. *The Destruction of Art: Iconoclasm and Vandalism since the French Revolution*. New Haven: Yale University Press, 1997.

Gary, Brett. *The Nervous Liberals: Propaganda Anxieties from World War I to the Cold War*. New York: Columbia University Press, 1999.

Geer, Galen L. "Nazi War Art." *Gung-Ho: The Magazine of the International Military Man*, October 1981, 34–59.

Genton, Bernard. *Les allies et la culture: Berlin, 1945–1949.* Paris: Presses Universitaires de France, 1998.

Gerhardt, Uta. "A Hidden Agenda of Recovery: The Psychiatric Conceptualization of Re-education for Germany in the United States during World War II." *German History* 14, no. 3 (1996): 297–324.

"German Attitudes on Collective Guilt." *Information Control Review,* April 26, 1947.

"German Reactions: Mills of Death." *Military Government Weekly Information Bulletin,* no. 30 (February 23, 1946): 14–15.

Geyer, Michael. "The Place of the Second World War in German Memory and History." *New German Critique,* no. 71 (Spring–Summer 1997): 5–40.

Gibbs, Jo. "State Department Art Classed as War Surplus." *Art Digest* 22, no. 17 (June 1, 1948): 7.

———. "State Department Shows 'Goodwill' Pictures." *Art Digest* 21, no. 13 (October 1, 1946): 13.

Gienow-Hecht, Jessica C. E. "Art Is Democracy and Democracy Is Art: Culture, Propaganda, and the *Neue Zeitung* in Germany, 1944–1947." *Diplomatic History* 23, no. 1 (Winter 1999): 21–43.

———. "Friends, Foes, or Reeducators? Feinbilder and Anticommunism in the U.S. Military Government in Germany, 1946–1953." In *Enemy Images in History,* ed Ragnhild Fiebig von Hase and Ursula Lehmkuhl, 281–301. Providence: Berghahn Books, 1997.

———. *Transmission Impossible: American Journalism and Cultural Diplomacy in Postwar Germany, 1945–1955.* Baton Rouge: Louisiana State University Press, 1999.

Geoffrey J. Giles, ed. *Archivists and Historians.* Washington, DC: German Historical Institute, 1996.

Gilkey, Gordon W. "German War Art." Office of the Chief Historian Headquarters, European Command, April 25, 1947.

Gillen, Eckart, and Diether Schmidt, eds. *Zone 5: Kunst in der Viersektorenstadt 1945–1951.* Berlin: Dirk Nishen, 1989.

Gimbel, John. *The American Occupation of Germany.* Stanford: Stanford University Press, 1968.

———. *A German Community under American Occupation: Marburg, 1945–1952.* Stanford: Stanford University Press, 1961.

Glaser, Hermann. *Deutsche Kultur: Ein historischer Überblick von 1945 bis zur Gegenwart.* Munich: Carl Hanser, 1997.

———. *The Rubble Years: The Cultural Roots of Postwar Germany, 1945–1948.* New York: Paragon House, 1986.

Goedde, Petra. "From Villains to Victims: Fraternization and the Feminization of Germany, 1945–1947." *Diplomatic History* 23, no. 1 (Winter 1999): 1–48.

Goldstein, Cora Sol. "Before the CIA: American Actions in the German Fine Arts (1946–1949)." *Diplomatic History* 29, no. 5 (November 2005): 747–78.

———. "The Control of Visual Representation: American Art Policy in Occupied

Germany, 1945–1949." *Intelligence and National Security* 18, no. 2 (June 2003): 283–99.

———. "Irak: Befreier in Ketten." *Internationale Politik* 60, no. 11 (November 2005): 104–10.

———. "Power and the Visual Domain: Images, Iconoclasm, and Indoctrination in American Occupied Germany, 1945–1949." PhD thesis, Department of Political Science, University of Chicago, 2002.

———. "The *Ulenspiegel* and Anti-American Discourse in the American Sector of Berlin." *German Politics and Society* 23, no. 2 (November 2005): 28–49.

Goodell, Stephen, and Susan D. Bachrach, eds. *Liberation 1945.* Washington, DC: United States Holocaust Memorial Museum, 1995.

"'Government' Art Called Mediocre." *New York Times,* March 9, 1952.

Green, Adwin Wigfall. *The Man Bilbo.* Baton Rouge: Louisiana State University Press, 1963.

Grosshans, Henry. *Hitler and the Artists.* New York: Holmes & Meier, 1983.

Groys, Boris. *The Total Art of Stalinism: Avant-Garde, Aesthetic Dictatorship, and Beyond.* Princeton: Princeton University Press, 1992.

Guilbaut, Serge. *How New York Stole the Idea of Modern Art: Abstract Expressionism, Freedom, and the Cold War.* Chicago: University of Chicago Press, 1983.

Gulgowski, Paul W. *The American Military Government of United States Occupied Zones of Post Worl War II Germany in Relation to Policies Expressed by Its Civilian Governmental Authorities at Home, during the Course of 1944/45 through 1949.* Frankfurt am Main: HAAG und Herchen, 1983.

Hahn, Brigitte J. *Umerziehung durch Dokumentarfilm? Ein Instrument amerikanischer Kulturpolitik im Nachkriegsdeutschland (1945–1953).* Munster: Lit Verlag, 1997.

Hammett, Ralph W. "Comzone and the Protection of Monuments in North-West Europe." *College Art Journal* 5, no. 2 (January 1946): 123–26.

Hartenian, Larry. *Controlling Information in U.S. Occupied Germany, 1945–1949: Media Manipulation and Propaganda.* Lewiston: Edwin Mellen Press, 2003.

Hartman, Geoffrey H., ed. *Holocaust Remembrance: The Shapes of Memory.* Oxford: Blackwell, 1994.

Haus der Kunst 1937–1997: Eine Historische Dokumentation. Munich: Haus der Kunst München, 1997.

Hauser, Johannes. *Neuaufbau der westdeutschen Filmwirtschaft 1945–1955.* Freiburg: Centaurus-Verlagsgesellschaft, 1989.

Heibel, Yule F. *Reconstructing the Subject: Modernist Painting in Western Germany, 1945–1950.* Princeton: Princeton University Press, 1995.

Hein-Kremer, Maritta. *Die amerikanische Kulturoffensive: Gründung und Entwicklung der amerikanischen Information-Centers in Westdeutschland und West-Berlin 1945–1955.* Cologne: Böhlau Verlag, 1996.

Heinzlmeier, Adolf. *Nachkriegsfilm und Nazifilm.* Frankfurt am Main: Oase Verlag, 1988.

Held, Jutta. *Kunst und Kunstpolitik 1945–1949*. Berlin: Elefanten Press, 1981.
"Henry Koerner." *Heute,* no. 36 (May 15, 1947), 18–19.
"Henry Koerner: His Own Tragedy Spurs Artists to Paint Moving Postwar Pictures." *Life,* May 10, 1948.
Herf, Jeffrey. *Divided Memory: The Nazi Past in the Two Germanys*. Cambridge: Harvard University Press, 1997.
Hilberg, Raul. *The Destruction of the European Jews*. New York: Holmes & Meier, 1985.
Hill, Gladwin. "Our Film Program in Germany: II. How Far Was It a Failure?" *Hollywood Quarterly* 2, no. 2 (January 1947): 131–37.
Hochgeschwender, Michael. "The Intellectual as Propagandist: *Der Monat,* the Congress of Cultural Freedom, and the Process of Westernization in Germany." Presented at "The American Impact on Western Europe: Americanization and Westernization in Transatlantic Perspective." Conference at the German Historical Institute, Washington, DC, March 25–27, 1999.
Hodin, J. P. "German Criticism of Modern Art since the War." *College Art Journal* 17, no. 4 (Summer 1958).
Hoenisch, Michael, Klaus Kämpfe, and Karl-Heinz Pütz. *USA und Deutschland: Amerikanische Kulturpolitik 1942–1949*. Berlin: J. F. Kennedy-Institut für Nordamerikastudien, 1980.
Hoffmann, Hilmar. *The Triumph of Propaganda: Film and National Socialism, 1933–1945*. Providence: Berghahn Books, 1996.
Hofman, Werner. *Wie deutsch ist die deutsche Kunst?* Leipzig: E. A. Seeman, 1999.
Honnef, Klaus, and Hans M Schmidt, eds. *Kunst und Kultur im Rheinland und in Westfalen 1945–1952: Aus den Trümmern—Neubeginn und Kontinuität*. Bonn: Rheinland Verlag, 1985.
Howley, Frank. *Berlin Command*. New York: G. P. Putnam's Sons, 1950.
Hutt, Wolfgang. *Deutsche Malerei und Graphik im 20. Jahrhundert*. Berlin: Henschelverlag Kunst und Gesellschaft, 1969.
Huyssen, Andreas. *Twighlight Memories: Marking Time in a Culture of Amnesia*. New York: Routledge, 1995.
"Indisputable Proof." *News Chronicle,* April 19, 1945.
International Biographical Dictionary of Central European Emigres, 1933–1945. Vol. 1. Munich: K. G. Saur, 1983.
Jaeger, Klaus, and Helmut Regel, eds. *Deutschland in Trümmern: Filmdokumente der Jahre 1945–1949*. Oberhausen: Karl Maria Laufen, 1976.
Jarausch, Konrad H. *After Hitler: Recivilizing Germans, 1945–1955*. Oxford: Oxford University Press, 2006.
Jendricke, Bernhard. *Die Nachkriegszeit im Spiegel der Satire: Die satirischen Zeitschriften Simpl und Wespennest in den Jahren 1946 bis 1950*. Frankfurt am Main: Peter Lang, 1982.
Joesten, Joachim. "The German Film Industry, 1945–1948." *New Germany Report,* no. 3 (May 1948), 3–16.

———. *Germany: What Now?* Chicago: Ziff-Davis, 1948.

Joseph, Robert. "Comment and Criticism." *Arts and Architecture* 62 (September 1946): 16.

———. "Confession and Reaffirmation." *Arts and Architecture* 63 (May 1946): 14.

———. "Film Program for Germany." *Arts and Architecture* 62 (July 1945): 16.

———. "The Germans See Their Concentration Camps." *Arts and Architecture* 63 (September 1946): 14.

———. "Our Film Program in Germany: I. How Far Was It a Success?" *Hollywood Quarterly* 2, no. 2 (January 1947): 122–30.

Judt, Tony. *Postwar: A History of Europe since 1945.* New York: Penguin, 2005.

Junker, Detlef, ed. *The United States and Germany in the Era of the Cold War, 1945–1990: A Handbook.* Vol. 1, *1945–1968.* Washington, DC: German Historical Institute; Cambridge, Cambridge University Press, 2004.

Kämpfe, Klaus. "Educating the Re-Educators: Films in the Morale-Program of the U.S. Army." Berlin: JFK Materialien, vol. 16, 1980.

Katz, Barry M. *Foreign Intelligence: Research and Analysis in the Office of Strategic Services, 1942–1945.* Cambridge: Harvard University Press, 1989.

Kent, George O. *Historians and Archivists: Essays in Modern German History and Archival Policy.* Fairfax, VA: George Mason University Press, 1991.

Kirstein, Lincoln. "Art in the Third Reich—Survey, 1945." *Magazine of Art* 38, no. 6 (October 1945): 223–41.

Knigge, Volkhard. "Opfer, Tat, Aufstieg: Vom Konzentrationslager Buchenwald zur Nationalen Mahn—und Gedenkstätte der DDR." In *Versteinertes Gedenken: Das Buchenwald Mahnmal von 1958.* Buchenwald: Stiftung Gedenkstatten Buchenwald und Mittelbau Dora, 1997.

Knipping, Franz, and Jacques Le Rider, eds. *Frankreichs Kulturpolitik in Deutschland, 1945–1950.* Tübingen: Attempto Verlag, 1987.

Koch, Steven. *Double Lives: Stalin, Willi Münzenberg and the Seduction of the Intellectuals.* New York: Enigma Books, 2004.

Koerner, Joseph Leo, ed. *Henry Koerner 1915–1991: Unheimliche Heimat.* Vienna: Österreiche Galerie Belvedere, 1997.

Koppes, Clayton R., and Gregory D. Black. *Hollywood Goes to War: How Politics, Profits, and Propaganda Shaped World War II Movies.* New York: Basic Books, 1987.

Kormann, John G. *U.S. Denazification Policy in Germany 1944–1950.* Historical Division Office of the Executive Secretary Office of the U.S. High Commissioner for Germany, 1952.

Kormendi, Andre. "What German Art Survives." *Art News,* October 1948, 41–43.

Kracauer, Siegfried. "Propaganda and the Nazi War Film." In *From Caligari to Hitler: A Psychological History of the German Film.* Princeton: Princeton University Press, 1947.

Krause, Markus. *Galerie Gerd Rosen: Die Avantgarde in Berlin 1945–1949.* Berlin: Ars Nicolai, 1995.

Krenn, Michael, L. *Fall-Out Shelters for the Human Spirit: American Art and the Cold War.* Chapel Hill: University of North Carolina Press, 2005.

Krippendorff, Ekkehart, ed. *The Role of the United States in the Reconstruction of Italy and West Germany, 1943–1949.* Berlin: J. F. Kennedy-Institut für Nordamerikastudien, 1981.

Krohn, Klaus-Dieter, Erwin Rotermund, Lutz Winckler, and Wulf Koepke, eds. *Aspekte der Künstlerischen Inneren Emigration 1933 bis 1945.* Munich: Gesellschaft für Exilforschung, 1994.

Kuby, Erich. *The Russians and Berlin, 1945.* London: Heinemann 1965.

KZ: Bildbericht aus fünf Konzentrationslagern. Amerikanischen Kriegsinformationsamt im Auftrag des Oberbefehlshabers der Alliierten Streitkräfte, Spring 1945.

Lammersdorf, Raimund. "The Question of Guilt, 1945–1947: German and American Answers." Presented at "The American Impact on Western Europe: Americanization and Westernization in Transatlantic Perspective." Conference at the German Historical Institute, Washington, DC, March 25–27, 1999.

Lang, Rudolf. "Hanuš Burger 75 Jahre." In *Studienkreis Rundfunk und Geschichte: Mitteilungen* 10, no. 3 (July 1, 1984).

Laswell, Harold D., Ralph D. Casey, and Bruce Lannes Smith, eds. *Propaganda and Promotional Activities.* Chicago: University of Chicago Press, 1946.

Laswell, Harold D., and Saul K. Padover. *Psychological Warfare.* New York: Foreign Policy Association, 1951.

Laurie, Clayton D. *The Propaganda Warriors: America's Crusade against Nazi Germany.* Lawrence: University Press of Kansas, 1996.

Leff, Laurel. "A Tragic 'Fight in the Family': The New York Times, Reform Judaism and the Holocaust." *American Jewish History* 88, no. 1 (March 2000): 3–51.

Lehman, Christopher P. "The New Black Animated Images of 1946." *Journal of Popular Film and Television* 29, no. 2 (Summer 2001): 74–81.

Lehmann-Haupt, Hellmut. *Art under a Dictatorship.* New York: Oxford University Press, 1954.

———. "Art under Totalitarianism." Presented at a symposium. Metropolitan Museum of Art, New York, March 1952.

———. "Behind the Iron Curtain." *Magazine of Art* 44 (March 1951): 83–88.

———. "German Museums at the Crossroads." *College Art Journal* 7, no. 2 (Winter 1947–1948): 121–27.

Lerner, Daniel. *Sykewar, Psychological Warfare against Germany.* New York: George W. Stewart, 1949.

———, ed. *Propaganda in War and Crisis.* New York: George W. Stewart, 1951.

Levi, Neil. "'Judge for Yourselves!'–The *Degenerate Art* Exhibition as Political Spectacle." *October* 85 (Summer 1998): 41–64.

Lewinski, Jorge. *The Camera at War: A History of War Photography from 1848 to the Present Day.* New York: Simon & Schuster, 1980.

Linebarger, Paul M. A. *Psychological Warfare.* Washington, DC: Combat Forces Press, 1948.

Linton, Leonard. "Kilroy Was Here." Unpublished manuscript. 1997.

Lipstadt, Deborah. *Beyond Belief: The American Press and the Coming of the Holocaust, 1933–1945.* New York: Free Press, 1993.

Liss, Andrea. *Trespassing through Shadows: Memory, Photography, and the Holocaust.* Minneapolis: University of Minnesota Press, 1998.

Littleton, Taylor D., and Maltby Sykes, eds. *Advancing American Art.* Tuscaloosa: University of Alabama Press, 1989.

Losson, Nicolas, and Annette Michelson. "Notes of the Images of the Camps." *October* 90 (Autumn 1999): 25–35.

Luckach, Joan M. *Hilla Rebay: In Search of the Spirit in Art.* New York: George Braziller, 1983.

Maase, Kaspar. "'Americanization,' 'Americanness' and 'Americanisms': Time for a Change in Perspective?" Presented at "The American Impact on Western Europe: Americanization and Westernization in Transatlantic Perspective." Conference at the German Historical Institute, Washington, DC, March 25–27, 1999.

MacGregor, Morris J., Jr. *Integration of the Armed Forces 1940–1965.* Washington, DC: Center of Military History, United States Army, 1985.

Mahoney, Haynes R. "Germany Goes to the Movies." *HICOG Information Bulletin,* January 1951, 13–16.

Maltin, Leonard. *Of Mice and Magic: A History of American Animated Cartoons.* New York: Plume, 1987.

Marcuse, Harold. "Nazi Crimes and Identity in West Germany: Collective Memories of the Dachau Concentration Camp, 1945–1990." PhD thesis, University of Michigan, 1992.

Margolin, Victor, ed. *Propaganda: The Art of Persuasion: World War II.* New York: Chelsea House, 1975.

Mathews, Jane de Hart. "Art and Politics in Cold War America." *American Historical Review* 81, no. 4 (October 1976): 762–87.

Mayer, Gerald M. "American Motion Pictures in World Trade." *Annals of the American Academy of Political and Social Science* 254 (November 1947): 31–36.

McBride, Joseph. *Frank Capra: The Catastrophe of Success.* New York: Simon & Schuster, 1992.

McClaskey, Beryl Rogers, ed. *Prolog 1: A Portfolio of Contemporary German Drawings and Prints Selected by a Group of American and German Residents of Berlin.* Berlin: Gebr. Mann, 1947.

McCreedy, Kenneth O. "Planning the Peace: Operation Eclipse and the Occupation of Germany." *Journal of Military History* 65, no. 3 (July 2001): 713–39.

McEvilley, Thomas. *The Exile's Return: Toward a Redefinition of Painting for the Post-Modern Era.* Cambridge: Cambridge University Press, 1993.

Medoff, Rafael. "'Our Leaders Cannot Be Moved': A Zionist Emissary's Reports on American Jewish Responses to the Holocaust in the Summer of 1943." *American Jewish History* 88, no. 1 (March 2000): 115–26.

Meehan, Patricia. *A Strange Enemy People: Germans under the British, 1945–1950.* London: Peter Owen, 2001.

Merritt, Anna J., and Richard L. Merritt. *Public Opinion in Occupied Germany: The OMGUS Surveys, 1945–1949.* Urbana: University of Illinois Press, 1970.

Merton, Robert K. *Mass Persuasion.* New York: Harper & Brothers, 1946.

Messaris, Paul. *Visual Persuasion: The Role of Images in Advertising.* Thousand Oaks: Sage Publications, 1997.

Michaud, Eric. *Un art de l'éternité: L'image et le temps du national socialisme.* Paris: Gallimard, 1996.

"Military Government Information Control." *Monthly Report of the Military Governor U.S. Zone,* January 1946.

"Military Government Information Control." *Monthly Report of the Military Governor U.S. Zone,* December–January 1947.

Millon, Henry A. and Linda Nochlin, eds. *Art and Architecture in the Service of Politics.* Cambridge: MIT Press, 1978.

Milton, Sybil, and Janet Blatter. *Art of the Holocaust.* New York: Routledge, 1981.

Mitchell, W. J. T., ed. *Art and the Public Sphere.* Chicago: University of Chicago Press, 1992.

Mnookin, Jennifer L. "The Image of Truth: Photographic Evidence and the Power of Analogy." *Yale Journal of Law and the Humanities* 10, no. 1 (Winter 1998): 1–74.

Moeller, Robert G. *War Stories: The Search for a Usable Past in the Federal Republic of Germany.* Berkeley: University of California Press, 2001.

Montgomery, John D. *Forced to Be Free: The Artificial Revolution in Germany and Japan.* Chicago: University of Chicago Press, 1957.

Morgan, Chester. *Redneck Liberal: Theodore G. Bilbo and the New Deal.* Baton Rouge: Louisiana State University Press, 1985.

Morgenthau, Hans, ed. *Germany and the Future of Europe.* Chicago: University of Chicago Press, 1951.

Morgenthau, Henry, Jr. *Germany Is Our Problem.* New York: Harper and Brothers, 1945.

Mosse, George L. *The Nationalization of the Masses: Political Symbolism and Mass Movements in Germany from the Napoleonic Wars through the Third Reich.* Ithaca: Cornell University Press, 1991.

———. *Nazi Culture.* New York: Grosset and Dunlap, 1966.

"Mr. Pulitzer's First Article from Paris." *St. Louis Post-Dispatch,* April 29, 1945.

Müller, Reinhard. *Menschenfalle Moscow: Exil und Stalinistische Vorfolgung.* Hamburg: Hamburger Edition, 2001.

Myers, D. G. "Jews without Memory: Sophie's Choice and the Ideology of Liberal Anti-Judaism." *American Jewish History* 88, no. 1 (March 2000): 499–529.

Naimark, Norman M. *The Russians in Germany: A History of the Soviet Zone of Occupation, 1945–1949.* Cambridge: Harvard University Press, 1995.

"Nazi Atrocity Films Real Shockers but U.S. Audiences Take It; Some Cuts." *Variety*, May 9, 1945, 17–18.

Nicholas, Lynn H. *The Rape of Europa*. New York: Vintage, 1995.

Nolan, Mary. "Americanization or Westernization?" Presented at "The American Impact on Western Europe: Americanization and Westernization in Transatlantic Perspective." Conference at the German Historical Institute, Washington, DC, March 25–27, 1999.

Novick, Peter. *The Holocaust in American Life*. Boston: Houghton Mifflin, 1999.

Occupation. United States Forces European Theater, 1946, 1–45.

Olick, Jeffrey K. *In the House of the Hangman: The Agonies of German Defeat, 1943–1949*. Chicago: University of Chicago Press, 2005.

Othman, Fred. "Scrambled Egg Art Sale." *Washington Daily News*, May 14, 1948.

Paddock, Alfred H., Jr. *U.S. Army Special Warfare: Its Origins: Psychological and Unconventional Warfare, 1941–1952*. Washington, DC: National Defense University Press, 1982.

Padover, Saul K. *Experiment in Germany: The Story of an American Intelligence Officer*. New York: Duell, Sloane and Pearce, 1946.

Pan, David. "The Struggle for Myth in the Nazi Period." *South Atlantic Review* 65, no. 1 (Winter 2000): 41–57.

Paret, Peter. *The Berlin Secession: Modernism and Its Enemies in Imperial Germany*. Cambridge: Belknap Press, 1980.

Parrish, Thomas. *Berlin in the Balance, 1945–1949: The Blockade, the Airlift, the First Major Battle of the Cold War*. Reading, Massachusetts: Perseus Books, 1998.

Pells, Richard. *Not Like Us: How Europeans Have Loved, Hated, and Transformed American Culture since World War II*. New York: Basic Books, 1997.

Petropoulos, Jonathan. *Art as Politics in the Third Reich*. Chapel Hill: University of North Carolina Press, 1996.

———. *The Faustian Bargain: The Art World in Nazi Germany*. Oxford: Oxford University Press, 2000.

Pfaff, Daniel W. *Joseph Pulitzer and the Post-Dispatch*. Pennsylvania: Pennsylvania State University Press, 1991.

"The Pictures Don't Lie." *Stars and Stripes*, April 26, 1945.

Pike, David. *The Politics of Culture in Soviet-Occupied Germany, 1945–1949*. Stanford: Stanford University Press, 1992.

Pilgert, Henry P. *The History of the Development of Information Services through Information Centers and Documentary Films*. Bonn: Historical Division Office of the Executive Secretary Office of the U.S. High Commissioner for Germany, 1951.

———. *Press, Radio and Film in West Germany, 1945–1953*. Bonn: Historical Division Office of the Executive Secretary Office of the U.S. High Commissioner for Germany, 1953.

Plischke, Elmer and Henry P. Pilgert. *U.S. Information Programs in Berlin*. Historical Division Office of the Executive Secretary Office of the U.S. High Commissioner for Germany, 1953.

Poiger, Uta G. *Jazz, Rock, and Rebels: Cold War Politics and American Culture in a Divided Germany.* Berkeley: University of California Press, 2000.

Pollock, James Kerr. *What Shall Be Done with Germany?* Northfield, MN: Carleton College, 1944.

Pommerin, Reiner. *The American Impact on Postwar Germany.* Providence: Berghahn Books, 1995.

Posey, Robert K. "Protection of Cultural Materials during Combat." *College Art Journal* 5, no. 2 (January 1946): 127–31.

Prolog: A Gift in Friendship for Beryl Rogers McClaskey and Charles Baldwin. Berlin: Brothers Hartmann, 1948.

Prolog. *A Report on German Museums.* Berlin, Fall 1947.

Pronay, Nicholas, and Keith Wilson, eds. *The Political Re-education of Germany and Her Allies after World War II.* London: Croom Helm, 1985.

"Protests to State Department Mount." *Art News* 46, no. 4 (June 1947): 8.

Pulitzer, Joseph. *A Report to the American People.* St. Louis: St. Louis Post-Dispatch, 1945.

Quigley, Martin. "Importance of the Entertainment Film." *Annals of the American Academy of Political and Social Science* 254 (November 1947): 65–69.

Randall, Richard S. *Censorship of the Movies: The Social and Political Control of a Mass Medium.* Madison: University of Wisconsin Press, 1968.

Ranke, Winfried, Carola Jüllig, Jürgen Reiche, and Dieter Vorsteher, eds. *Kultur, Pajoks und Care-Packete: Eine Berliner Chronik 1945–1949.* Berlin: Nishen, 1990.

Read, Herbert, "Will Grohmann." *Burlington Magazine* 110, no. 784 (July 1968): 413.

Redslob, Edwin. "Übersteigerte Wirklichkeit: Zur Ausstellung des amerikanischen Malers Henry Koerner." *Tagespiegel,* April 4, 1947.

Rentschler, Eric. *Ministry of Illusion: Nazi Cinema and Its Afterlife.* Harvard: Harvard University Press, 1996.

"Report from Germany: U.S. Films in Reopened Theaters." *Film News* 7, no. 3 (November–December 1945).

"Report on German Murder Mills." *Army Talks* 4, no. 9 (July 10, 1945).

"Report on Inspection of German Concentration Camp at Buchenwald." *Monthly Intelligence Summary for Psychological Warfare: How Much Do the Germans Know?* April 30, 1945.

Robb, Marilyn. "Chicago." *Art News* 46, no. 11 (January 1948): 39.

Robert, Philippe. *Political Graphics: Art as a Weapon.* New York: Abbeville Press, 1982.

Roditi, Edouard. "The Destruction of the Berlin Museums." *Magazine of Art* 42, no. 8 (December 1949): 306–12.

Roeder, George H. *The Censored War: American Visual Experience during World War Two.* New Haven: Yale University Press, 1995.

Roh, Franz. *German Art in the 20th Century.* New York: New York Graphic Society, 1968.

Rohde, Georg, ed. *Edwin Redslob zum 70. Geburstag: Eine Festgabe.* Berlin: Erich-Blaschker, 1955.

Rosenfeld, Frances A. "The Anglo-German Encounter in Occupied Hamburg, 1945–1950." PhD thesis, Columbia University, 2006.

Rosenfeld, Gavriel D. *Munich and Memory: Architecture, Monuments, and the Legacy of the Third Reich.* Berkeley: University of California Press, 2000.

Ross, Marvin C. "SHAEF and the Protection of Monuments in Northwestern Europe." *College Art Journal* 5, no. 2 (January 1946): 119–22.

Rottenburg, Bernd. *Das Haus am Waldsee unter der Leitung von L. K. Skutsch in den Jahren 1946 bis 1958.* Berlin, Diplomarbeit, Fachhochschule für Technick und Wirtschaft, 1997.

Ruby, Sigrid. "The Give and Take of American Painting in Postwar Western Europe." Presented at "The American Impact on Western Europe: Americanization and Westernization in Transatlantic Perspective." Conference at the German Historical Institute, Washington, DC, March 25–27, 1999.

———. *Have We an American Art?* Weimar: Verlag und Datenbank für Geisteswissenschaften, 1999.

Ruckert, Claudia, and Sven Kuhrau, eds. *Der Deutschen Kunst: Nationalgalerie und Nationale Identitat 1876–1998.* Amsterdam: Verlag der Kunst, 1998.

Sandberg, Herbert. *Frühe Karikaturen, Späte Graphik.* Berlin: Akademie der Künste der Deutschen Demokratische Republik, 1988.

———. *Herbert Sandberg: Leben und Werk.* East Berlin: Henschelverlag, 1977.

———. *Herbert Sandberg: 40 Jahre Graphik und Satire, Austellung zum 60 Geburstag.* East Berlin: Staatlichen Museum Berlin, 1968.

———. *Spiegel Eines Lebens.* East Berlin: Aufbau Verlag, 1988.

———. *Ulenspiegel: Deutschland vor der Teilung.* Oberhausen: Ludwig Institut Schloß Oberhausen, 1990s.

Sandberg, Herbert, and Gunter Kunert, eds. *Ulenspiegel: Zeitschrift fur Lieratur, Kunst und Satire1945–1950.* Berlin: Ulenspiegel Verlag, 1988.

Saunders, Frances Stonor. *The Cultural Cold War: The CIA and the World of Arts and Letters.* New York: New Press, 2000.

Schissler, Hanna, ed. *The Miracle Years: A Cultural History of West Germany, 1949–1968.* Princeton: Princeton University Press, 2001.

Schivelbusch, Wolfgang. *In a Cold Crater: Cultural and Intellectual Life in Berlin, 1945–1948.* Berkeley: University of California Press, 1998.

Schneede, Uwe. "Verisme et Nouvelle Objectivité." In *Paris-Berlin.* Paris: Gallimard, 1978.

Schulberg, Stuart. "Of All People." *Hollywood Quarterly* 4, no. 2 (Winter 1949): 206–8.

Schulte-Sasse, Linda. *Entertaining the Third Reich: Illusions of Wholeness in Nazi Cinema.* Durham: Duke University Press, 1996.

Schwartz, Thomas Alan. *America's Germany: John F. McCloy and the Federal Republic of Germany.* Cambridge: Harvard University Press, 1991.

Scott-Smith, Giles. "A Radical Democratic Political Offensive: Melvin J. Lasky, Der Monat, and the Congress for Cultural Freedom." *Journal of Contemporary History* 35, no. 2 (April 2000): 263–80.

Scott-Smith, Giles, and Hans Krabbendam, eds. *The Cultural Cold War in Western Europe*. London: Frank Cass, 2003.

Shain, Russel Earl. "An Analysis of Motion Pictures about War Released by the American Film Industry, 1939–1970." PhD thesis, University of Illinois, Urbana, 1971.

Shandley, Robert R. *Rubble Films: German Cinema in the Shadow of the Third Reich*. Philadelphia: Temple University Press, 2001.

Sheehan, James J. *Museums in the German Art World: From the End of the Old Regime to the Rise of Modernism*. Oxford: Oxford University Press, 2000.

Shore, Marci. *Caviar and Ashes: A Warsaw Generation's Life and Death in Marxism, 1918–1968*. New Haven: Yale University Press, 2006.

"Short Memories and Nice People." *Yank: The Army Weekly,* October 26, 1945.

Smith, Bradley F. *The Shadow Warriors: O.S.S. and the Origins of the C.I.A.* New York: Basic Books, 1983.

Smith, Tony. *America's Mission: The United States and the Worldwide Struggle for Democracy in the Twentieth Century*. Princeton: Princeton University Press, 1994.

Solomon, Charles. *The History of Animation: Enchanted Drawings*. New York: Wings Books, 1994.

Sontag, Susan. *On Photography*. New York: Picador, 2001.

———. *Regarding the Pain of Others*. New York: Picador, 2003.

Spotts, Frederic. *Hitler and the Power of Aesthetics*. Woodstock: Overlook Press, 2002.

Standen, Edith Appleton. "Report on Germany." *College Art Journal* 7, no. 3 (Spring 1948).

Stavitsky, Gail. *Henry Koerner: From Vienna to Pittsburgh*. Pittsburgh: University of Pittsburgh Press, 1983.

———. "Of Two Worlds: The Painting of Henry Koerner." *Arts Magazine* 60 (May 1986).

Stephan, Alexander, ed. *Americanization and Anti-Americanism: The German Encounter with American Culture after 1945*. New York: Berghahn Books, 2005.

———, ed. *The Americanization of Europe: Culture, Diplomacy, and Anti-Americanism after 1945*. New York: Berghahn Books, 2006.

Stocking, George W., Jr., ed. *A Franz Boas Reader: The Shaping of American Anthropology, 1883–1911*. Chicago: University of Chicago Press, 1974.

Strong, John W., ed. *McCormick on Evidence*. 5th ed. Vol. 2. St. Paul: West Group, 1999.

Stürickow, Regina. *Der Insulaner verliert die Ruhe nicht: Günter Neumann und seine Kabarett zwischen Kaltem Krieg und Wirtschaftwunder*. Berlin: Arani, 1993.

Sussex, Elizabeth. "The Fate of F3080." *Sight & Sound* 53, no. 2 (Spring 1984): 92–105.

Sweeney, Michael S. *Secrets of Victory: The Office of Censorship and the American Press and Radio in World War II*. Chapel Hill: University of North Carolina Press, 2001.

Szluk, Peter F. "'Freedom' Prizes for Artists." *HICOG Information Bulletin*, June 1950, 33–35.

Tabor, Jan, ed. *Kunst und Diktatur: Architektur, Bildhauerei und Malerei in Österreich, Deutschland, Italien un der Sowjetunion 1922–1956*. Baden: Verlag Grasl, 1994.

Taylor, Brandon, and Wilfried van der Will. *The Nazification of Art: Art, Design, Music, Architecture, and Film in the Third Reich*. Winchester: Winchester Press, 1990.

Tegel, Susan. "Veit Harlan and the Origins of 'Jud Süss,' 1938–1939: Opportunism in the Creation of Nazi and Anti-Semitic Film Propaganda." *Historical Journal of Film, Radio, and Television* 16, no. 4 (1996): 515–32.

Tent, James F. *Mission on the Rhine: Reeducation and Denazification in American-Occupied Germany*. Chicago: University of Chicago Press, 1982.

Thompson, George R., and Dixie R. Harris. *The Signal Corps*. Washington, DC: Office of the Chief of Military History, United States Army, 1966.

Thomson, Charles A. H. *Overseas Information Service of the United States Government*. Washington, DC: Brookings Institution, 1948.

Thomson, Charles A. H., and Walter H. C. Laves. *Cultural Relations and U.S. Foreign Policy*. Bloomington: Indiana University Press, 1963.

"Torture Camps: This Is the Evidence." *Daily Mail*, April 19, 1945.

Tuch, Hans N. *Communicating with the World: U.S. Public Diplomacy Overseas*. New York: St. Martin's Press, 1990.

Turner, Jane, ed. *The Dictionary of Art*. Willard: Macmillan, 1996.

Ullman, Frederic. "Russia Eclipses U.S. on Pix in Reich." *Variety*, 1947.

U.S. Department of State. *Germany 1945–1949: The Story in Documents*. Washington, DC: U.S. Government Printing Office, 1950.

Van Loon, Gerard W. "The Theater Returns to the Bavarian Scene." *Military Government Weekly Information Bulletin*, no. 26, January 26, 1946.

van Sweringen, Brian T. *Kabarettist an der Front des Kalten Krieges: Günter Neumann und das politische Kabarett in der Programmgestaltung des Radios im amerikanischen Sektor Berlins (RIAS)*. Passau: Richard Rothe, 1989.

Virilio, Paul. *Guerre et cinéma: Logistique de la perception*. Paris: Cahiers du Cinema, 1991.

von Hippel, Karin. *Democracy by Force: U.S. Military Intervention in the Post-Cold War World*. Cambridge: Cambridge University Press, 2000.

Wade, Warren. "How Is Combat Footage Used?" *Business Screen* 7, no. 1 (January 1948).

Wagnleitner, Reinhold. *Coca-Colonization and the Cold War: The Cultural Mission of the United States in Austria after the Second World War*. Chapel Hill: University of North Carolina Press, 1994.

"War Department Releases Atrocity Films for Nation to See." *St. Louis Post-Dispatch,* June 7, 1945.

Warner, Michael. *The Office of Strategic Services: America's First Intelligence Agency.* Washington, DC: CIA Public Affairs Office, 2000.

Washton-Long, Rose-Carol. *German Expressionism: Documents from the End of the Wilhelmine Empire to the Rise of National Socialism.* New York: G. K. Hall, 1993.

Watson, Mark Skinner. *United States Army in World War II.* Washington, DC: Center of Military History, United States Army, 1991.

Weber, John Paul. *The German War Artists.* Columbia, SC: Cerberus, 1979.

Weidler, Charlotte. "Art in Western Germany Today." *Magazine of Art* 44, no. 4 (1951): 132–38.

Weisenborn, Günther. *Der gespaltene Horizont.* Munich: Desch, 1964.

Weisgall, Hugo, ed. *Advancing American Art 1947.* Prague: U.S. Information Service, 1947.

Welch, David. *Propaganda and the German Cinema.* Oxford: Clarendon Press, 1983.

———. *The Third Reich: Politics and Propaganda.* London: Routledge, 1995.

Whiting, Charles. *Patton's Last Battle.* New York: Stein and Day, 1987.

Wiesen, S. Jonathan. *West German Industry and the Challenge of the Nazi Past, 1945–1955.* Chapel Hill: University of North Carolina Press, 2001.

Willett, Ralph. *The Americanization of Germany, 1945–1949.* London: Routledge, 1989.

———. "Billy Wilder's 'A Foreign Affair' (1945–1948): The Trials and Tribulations of Berlin." *Historical Journal of Film, Radio, and Television* 7, no. 1 (1987): 3–14.

Willoughby, John. *Remaking the Conquering Heroes: The Social and Geopolitical Impact of the Post-War American Occupation of Germany.* New York: Palgrave, 2001.

Winkler, Allan. *The Politics of Propaganda: The Office of War Information, 1942–1945.* New Haven: Yale University Press, 1978.

Wolfe, Robert, ed. *Americans as Proconsuls: United States Military Government in Germany and Japan, 1944–1952.* Carbondale: Southern Illinois University Press, 1984.

Wyman, David S. *The Abandonment of the Jews: America and the Holocaust, 1941–1945.* New York: Pantheon, 1984.

———. *America and the Holocaust.* New York: Garland Publishing, 1990.

Zelizer, Barbie. *Remembering to Forget: Holocaust Memory through the Camera's Eye.* Chicago: University of Chicago Press, 1998.

———, ed. *Visual Culture and the Holocaust.* New Brunswick: Rutgers University Press, 2000.

Zielinski, Siegfried, and Gloria Cunstance. "History as Entertainment and Provocation." *New German Critique* 19, no. 1 (Winter 1980): 81–96.

Zimke, Earl. *The U.S. Army Occupation of Germany, 1944–1946.* Washington, DC: Center of Military History, U.S. Army, 1975.

Zink, Harold. *American Military Government in Germany.* New York: MacMillan, 1947.

———. *The United States in Germany, 1944–1955.* Princeton: D. Van Nostrand, 1957.

Zolotow, Maurice. *Billy Wilder in Hollywood.* Pompton Plains: Limelight Editions, 2004.

Index

Adams, Paul, 31–32
Adenauer, Konrad, 123, 130, 131
Advancing American Art (exhibit, 1946–1947), 78–80
African American soldiers, 61–62, 154n91
Agee, James, 43
Allgemeine Deutsche Kunst Ausstellung ("German Art Exhibit," 1946), 77
Allied Control Council, 10, 106, 138n22
Almabtrieb (Wuellfarth), 100
Altenbourg, Gerhard, 162n23
American Artists Professional League, 79
American Commission for the Protection and Salvage of Artistic and Historic Monuments (Roberts Commission), 75
American Commission on Public Information (Creel Committee), 105
American intelligence: evolution of, 5–9; Military Intelligence Division, 6–7; Office of Strategic Services, 6–8; Truman authorizes to investigate federal employees, 16, 116. *See also* Central Intelligence Agency (CIA)
Amerika Dienst (U.S. Feature Service), 121
Amerikahäuser (United States Information Centers), 99, 102, 119, 163n34
Amery, Leopold, 42
amnesty, 122
Amtliches Material zum Massenmord von Katyn (propaganda booklet), 34
anticommunism: American propaganda becomes focused on, 17–18; and *The Brotherhood of Man*, 67; of *Der Insulaner*, 121; *Kultur* associated with, 101; of Military Intelligence Division, 6; OMGUS anticommunist campaign, 4, 17, 116, 128, 130–31; Prolog group associated with, 91–92; Republicans use against New Dealers, 5; Truman's ideological war against communism, 16–17; West Germans seen as allies in, 38. *See also* Cold War
anti-immigration movement, 6
anti-Semitism: as absent in wartime propaganda, 23; American documentary films for export ignore, 46; of Bilbo, 152n78; in Goebbels's anti-Soviet propaganda campaign, 34; of Military Intelligence Division, 6; in Nazi propaganda films, 47; persists in West Germany, 131; Soviet anti-Semitic offensive, 16, 123, 124; *Todesmühlen* does not comment on, 55
antislavery propaganda, 27
"Architecture in the Third Reich" (Kallweit), 113
art. *See* fine arts
Art Digest (magazine), 93
Art Intelligence Unit, 94
Art News (magazine), 79
Art under Dictatorship (Lehmann-Haupt), 159n43
"Assault on German Culture" (Schlichter), 113

Atrocities: A Study of German Reactions (Psychological Warfare Division), 33, 34, 145n6
atrocity propaganda, 21–39; effect on American home front, 22–28; effect in Germany, 28–39; evolution of, 35–39; film in, 21–22; initial impact in Germany, 32–35; *Todesmühlen,* 21–22, 51–57; as visual propaganda, 3–4, 128, 129
Augstein, Rudolf, 19
Ausstellung: Henry Koerner U.S.A. 1945–1947 (exhibit), 92–95, 161n13
Autobiography of a Jeep (film), 49

Backes, Johannes, 162n23
Barkley, Alben, 26, 141n17
Barlach, Ernst, 77
Barr, Alfred, 76, 159n43
Battle of Britain (film), 45
Battle of China (film), 45
Battle of Russia (film), 45
Baumeister, Willi, 98, 99, 163n33
Becher, Johannes R., 111, 168n23
Beckmann, Max, 77
Béguin, Albert, 20
Benedict, Ruth, 7, 62–63, 63–64, 67
Benton, William, 78, 80
Berlin: destruction in, 9, 138n21; division of, 10; movie attendance in American sector of, 56; *Ulenspiegel*'s perception of political situation in, 114; *Zwischen West und Ost* depicts, 60
Berlin Blockade and Airlift, 58, 59–60, 91, 95, 130
Berliner Blatt (newspaper), 119
Berliner Zeitung (newspaper), 116
Berlin Secession Movement, 71
Bernstein, Sidney L., 52, 148n41
Bilbo, Theodore Gilmore, 63–64, 152n78
Bildende Kunst (journal), 125
Bizonia, 115
Black, Percy, 8

Blaue Reiter, 71
Blevins Davis Prize, 99–100
Boas, Franz, 62, 63–64, 152n78
Bohrer, Harry, 19
Bohringen, Volker, 162n23
Bormann, Martin, 72, 107
Boswell, Helen, 93
Bourke-White, Margaret, 29
Bradley, Omar, 21, 22, 35
Brandt, Willy, 132
Braun, Eduard, 110
Brecht, Bertolt, 111, 112
Breen, Robert, 100
Breusing, Ima, 161n7
Britain. *See* Great Britain
British Council Branches, 19
Brotherhood of Man, The (film), 61–67; as based on Benedict and Weltfish's *The Races of Mankind,* 62–63, 65, 67; communist associations of makers of, 67; controversy about showing in occupied Germany, 65–66, 129; production of, 62, 64
Brücke, Die (British cultural program), 19
Brücke, Die (film), 59–60
Brücke, Die (modern art movement), 71
Bruller, Jean-Marcel (Vercors), 20
Buchholz, Karl, 96
Büchner, Robert, 161n13
Burger, Hanus, 52–53, 55
burials, public, 31–32
Busbey, Fred, 79–80, 158n35
Byrnes, James F., 115

Cain, Harry, 32
Cannon, Bob, 64
Capra, Frank, 26, 45
Carlebach, Emil, 111, 117, 168n22
Catholic Legion of Decency, 43, 105
censorship, 105–26; Cold War and imposition of, 126; expands to include communism, 17, 18; German social-

ists denounce imperial, 71; by Information Control Division, 12–13; of modern art by Nazis, 73; of war photography, 27, 28
Central Intelligence Agency (CIA): covert operations of, 97; cultural modus operandi of, 89, 160n1; Information Control Division as influence on, 133; and *Der Monat*, 121; Office of Strategic Services' model adopted by, 8; responsibilities of, 5–6
Chagall, Marc, 20, 112
Challenge to Democracy (film), 46
Chaloner, John, 19
Chief of Staff of the Supreme Allied Commander (COSSAC), 137n16
Christian Democratic Union, 131–32
Churchill, Winston, 106, 115
CIA. *See* Central Intelligence Agency (CIA)
cinema. *See* film
Civil Affairs Division (CAD), 41
Clay, Lucius D.: on arts policy, 86; on black soldiers, 61; Friedrich as advisor to, 11; Lasky becomes consultant to, 17; local elections held by, 12; as not interested in cultural affairs, 13; Operation Talk Back launched by, 17, 117; and Prolog project, 90; tasks as military governor, 10; *Ulenspiegel* on, 123, 124
Cohen, Emanuel, 45
Cold War: American policy in occupied Germany affected by, 38; British cultural policy affected by, 19; denazification policy affected by, 122; Documentary Film Unit propaganda films, 58, 59–60; gives German artists bargaining power with OMGUS, 95–97; OMGUS press policy affected by, 115–17, 126, 128–29; OMGUS propaganda changes in response to, 17–18; psychological warfare in, 1;

race as issue in anti-American propaganda in, 62, 151n71
collective guilt: American atrocity propaganda on, 21, 28, 30, 32, 34; American propaganda moves away from doctrine of, 35, 36–37, 130; Cold War and deactivation of doctrine of, 38; Germans reject doctrine of, 39; Soviets reject doctrine of, 34; in *Todesmühlen*, 51, 55, 56
Colleges at War (film), 46
concentration camps: in American atrocity propaganda, 3–4, 21–39; congressional delegation to, 24–25, 141n17; disappear from American propaganda, 38–39; editors tour, 25, 142n18; effect on American home front, 22–28; effect on Germany, 28–39; end of visits to, 35; forced visits to, 21, 30–31, 35; and German industry, 53; as justification for American involvement in war and occupation, 132; liberation of, 140n8; Nazi films ignore, 47; photographs of, 21, 24, 25–26, 39, 51; smells and sights of, 22; in *Todesmühlen*, 21–22, 51–57; in *Ulenspiegel* iconography, 169n27
confrontation policy: as failure, 32–33, 39, 130; forced tours of the camps, 21, 30; Soviet propaganda contrasted with, 34, 35; successes of, 130; *Todesmühlen* documentary as element of, 51, 52; visual component of, 21–22, 129
Constable, William G., 102
constitutional patriotism, 132
containment policy, 116
Control Council Directive No. 30, 106
"Conversation on the Border" (Sandberg), 118
cosmopolitanism, 16, 71, 73, 77, 78, 123, 124

Creel Committee (American Commission on Public Information), 105
Crew, David, 39
Crimean War, 27
Crisis (film), 52
Crist, W. E., 64
culture: Americans arrive without cultural agenda, 13, 129; cultural initiatives in British zone, 18–19; cultural initiatives in French zone, 19–20; *Kultur* promoted by Americans, 101–3; as political weapon, 2; in Soviet-occupied Germany, 4–5, 13–16. *See also* fine arts

Daniell, Raymond, 65
Darre, Walter, 72
Davidson, J. Leroy, 78, 80
Davis, Blevins, 99–100, 164n37
Davis, Clifford, 63
Davis, Elmer, 3, 44, 51
Davis, Stuart, 78
De Gaulle, Charles, 123
"Degenerate Art Exhibit" (*Entartete Kunst Ausstellung*, 1937), 3, 73, 74, 77
Degeneration (Nordau), 70
Delarbre, Léon, 162n22
de Mendelssohn, Peter, 111
Democracy in Action (film), 49
democratization, 10, 11, 12, 13–14, 41, 127
denazification: British concern about American model of, 19; Cold War alters American policy of, 122; versus collective guilt, 37; of German press, 108–9; information control in, 12; Lehmann-Haupt on rejection of modern art and, 85; Nazi iconography and paraphernalia eliminated, 106, 128; in OMGUS mission, 10; in Soviet zone, 123; *Ulenspiegel*'s criticism of, 113, 122, 129; variation in American views regarding, 11, 127; visual propaganda in, 2; of visual sphere, 127; Wilder as consultant on, 53
desegregation of the military, 61–62, 66, 154n91
Deutsche Verwaltung für Volksbildung, 15, 76
DFU. *See* Documentary Film Unit (DFU)
Diese Woche (magazine), 19
Divide and Conquer (film), 45
Dix, Otto, 71, 77, 93, 100
Döblin, Alfred, 113
Doctors at War (film), 46
documentary films: in American film policy in occupied Germany, 49–51; as interpretive, 42; as propaganda, 42–44; *Todesmühlen*, 21–22, 51–57. *See also* Documentary Film Unit (DFU)
Documentary Film Unit (DFU), 57–60; *Die Brücke*, 59–60; Cold War in establishment of, 18; *Hunger*, 58–59; two types of film of, 58; *Zwei Städte*, 60; *Zwischen West und Ost*, 60
Dondero, George, 79, 158n33
Donovan, William "Wild Bill," 7
Draper, William H., Jr., 66, 86–87
Durham, Carl T., 63
Dymschitz, Alexander L., 14, 15, 16, 77, 81, 83–84, 124–25

Eastman, Phil, 64, 67
Eberle, George E., 66
Effel, Jean, 118
Ehmsen, Heinrich, 113, 162n23
Eisenhower, Dwight D.: concentration camps visited by, 21, 22, 55; on exposing atrocity propaganda on home front, 24, 39, 141n16; as not particularly interested in cultural affairs, 13; on protecting cultural treasures, 75; stops visits to concentration camps, 35
Elliot, John, 169n34

Entartete Kunst Ausstellung ("Degenerate Art Exhibit," 1937), 3, 73, 74, 77
Erhard, Ludwig, 131
eugenics movement, 6
European Theater of Operation, U.S. Army (ETOUSA), 137n17
"euthanasia" films, 47
Evans, Luther, 26
Evans, Walker, 43
Ewige Jude, Der (film), 47
Executive Order 9835, 16, 116
Executive Order 9981, 154n91
Expressionism, German, 72
extermination camps. *See* concentration camps

Fahy Committee, 154n91
Farlow, Arthur C., 62–63
Federal Employee Loyalty Program, 16
Feininger, Lyonel, 77
Felixmüller, Conrad, 77
Fenton, Roger, 27
"Fifty-Fifty" (Sandberg), 118–19
Fight for Life, The (film), 43
film: American film-going audience, 44; American film policy in occupied Germany, 46–51; American propaganda films, 41–67; in atrocity propaganda, 21–22; censorship of, 12; German film attendance, 46; importance as medium of mass indoctrination, 42; *Im Wald von Katyn*, 34; in *Lest We Forget* exhibit, 26; *The Nazi Concentration Camps* at Nuremberg trials, 36; Nazi propaganda films, 46–47; OMGUS employs, 4, 13, 41, 47–51, 128; of Political Intelligence Division of the Foreign Office, 19; as political reeducation, 48–49; Soviet use of, 15, 47–48. *See also* Hollywood films; nonfiction films
fine arts, 69–103; American policy regarding, 4–5, 74; censorship of, 12–13; Christian thematics in postwar German, 100; Cold War gives German artists bargaining power with OMGUS, 95–97; French military government sponsorship of, 20; in Germany, 70–74; Hitler as obsessed with, 71; in iconoclasm campaign, 107; immediate postwar German, 94, 162n22, 162n23; Information Control Division and German, 74–76; as Information Control Division's blind spot, 69–87; *Kultur* and American policy toward, 101–3; Marxist politicization of, 125; Monuments, Fine Arts & Archives Section in arts policy, 69–70, 81–87; Nazis' positive policy for, 73–74; Nazi stolen art, 69, 75, 87, 107; overt and covert American actions in German, 89–103; Prolog group, 89–92; as propaganda, 69, 83, 97; Socialist Realism, 5, 77, 83, 125; Soviet policy regarding, 15, 16, 69, 76, 97; in Soviet zone and sector, 76–77. *See also* modern art
Firetz, Gerhard, 98–99
First German Writers' Congress, 17, 111
Fleischer, Jack, 169n34
Fodor, Marcel W., 169n34
folklore, 167n14
Folkwang Museum (Essen), 71
Ford, John, 57
Foreign Affair, A (film), 54
"Formalism and Realism in the Fine Arts" (debate), 125
Foss, Kendall, 169n34
Four D's, 10
Fragebogen, 109, 113
France: Berlin sector of, 10; cultural initiatives in French zone, 19–20; zone of occupation of, 10
Frankenstein, Wolfgang, 98
Frankfurter Rundschau (newspaper), 111, 117
fraternization, 22, 23, 35, 36–37

"Freedom and Equality" (Sandberg), 118
Free Trade Union (FDGB), 95, 96
Frick, Wilhelm, 72
Friedrich, Carl J., 11
Frischen Wind (journal), 123, 125
Fulbright Act, 78
Fulton, Missouri, speech, 115
Furet, François, 122

Gaedicke, Hans, 96
Gavin, James, 30
Geiger, Rupprecht, 98
Geiger, Willi, 162n23
Geisel, Theodore Seuss, 45
General Staff Divisions, 137n15
Geography of Japan (film), 46
George, Walter, 26, 141n17
"German Art Exhibit" (*Allgemeine Deutsche Kunst Ausstellung*, 1946), 77
"German Art Today" (Lehmann-Haupt), 85
"German Buildings, The" (Sandberg), 114
German Expressionism, 72
Germany: conditions at end of war, 9–10; fine arts in, 70–74. See also Berlin; Nazis; occupied Germany
Gesellschaft für Christliche Kultur, 100
Gibbs, Jo, 93
Gilkey, Gordon W., 107
Gillem, Alvan C., Jr., 154n91
Gimbel, John, 1–2
Goebbels, Joseph, 34, 46, 72, 74
Gone with the Wind (film), 49
Göring, Hermann, 59, 72
Government Information Manual for the Motion Picture Industry, 44–45
Grace, Alonzo, 90
Granland, Ed, 29
Grapes of Wrath, The (film), 49
Great Britain: Berlin sector of, 10; cultural initiatives in British zone, 18–19; as depicted in American films, 45; film employed in occupied Germany by, 47, 48; nonfiction film for propaganda in, 42–43; and United States consolidate their zones, 115; zone of occupation of, 10
"Great German Art Exhibit" (*Große Deutsche Kunstausstellung*), 74
Grierson, John, 42, 46
Grochowiak, Thomas, 100
Grohmann, Will, 96
Grosz, George, 71, 78, 93, 110, 112, 163n33
Grundig, Hans, 162n23
Grundig, Lea, 162n23
Guggenheim Foundation, 98, 99
Günter Neumann und seine Insulaner (radio program), 121

Habe, Hans (János Bekéssy), 115–16
Haftmann, Werner, 99
Hall, John A., 25
Hamlet (Shakespeare), 100
Harmon, Ernest, 61
Harris, Ken, 64
Hartmann, Kurt, 91, 96
Hartung, Karl, 161n7
Harvard Group, 75
Harvard School of Overseas Administration, 11
Hassett, William D., 65
Haus am Waldsee, 92, 161n13
Haus der Kunst (Munich), 74
Hays, William, 43, 145n6
Hearst newspapers, 78, 79
Hebbel Theater, 111
Heckel, Erich, 77
Hegenbarth, Josef, 161n7
Heiliger, Bernard, 161n7
Heineman, J. Paul, 29
Heinrich, Theodore A., 97
Held, Jutta, 162n22
Hempel, Willi, 99

Henselmann, Hermann, 113
Here's Your Army (war show), 164n37
Hermand, Jost, 162n22
"Heuss: Take Courage and Jump into Independence" (Sandberg), 124
Heute (magazine), 93–94
Heymann, Stefan, 125
HICOG Information Bulletin, 91–92
HICOG Monthly Bulletin, 39
High Commissioner for Germany (HICOG), 18, 127
Hill, Gladwin, 48, 50
Himmler, Heinrich, 107
Hine, Lewis, 27
Hitler, Adolf: art collection of, 107; as arts patron of Third Reich, 73–74; modern art rejected by, 72; as obsessed with art and architecture, 71; *Ulenspiegel*'s depictions of, 113
Hofer, Karl, 91, 112, 113, 161n7
Hoffmann, Heinrich, 72
Hollywood films: American conservatives employ, 81; censorship of, 43, 105; Germany as prewar competitor of, 46; *L'Humanité* on, 151n71; overseas revenues of, 146n15; political directives for wartime films, 44–45; in reeducation program for occupied Germany, 41, 48, 49; United States Film Service as competitor of, 43; in war effort, 3; Wilder's importance to, 54
Holocaust: *Ausstellung: Henry Koerner U.S.A. 1945–1947* exhibit, 92–95; first widely distributed images of, 39; Germany still affected by, 133; *Hunger* does not discuss, 58; *Todesmühlen* ignores Jewish specificity of, 55; Wilder distances himself from, 54. *See also* concentration camps
Holocaust (miniseries), 132
Holtz, Karl, 110, 119
Hoover, J. Edgar, 6

House Un-American Activities Committee (HUAC), 67, 80
Howard, Richard, 83, 86, 90, 92, 93, 161n13
Hubley, John, 64, 67, 153n81
Huggler, Max, 99
Hull, Cordell, 11
Humanité, L' (newspaper), 151n71
Hunger (film), 58–59
Huston, John, 49
Hutton, Tom, 95, 96, 97

ICD. *See* Information Control Division (ICD)
iconoclasm: OMGUS campaign of, 4, 105, 106–8, 128; radical regime change accompanied by, 105; Ziegler's campaign against modern art, 73
Illegalen, Die (Weisenborn), 111
Illustrated (magazine), 25
Illustrated London News, 24
Im Wald von Katyn (film), 34
incentive films, 45–46, 146n13
Industrial Britain (documentary film), 42
information control: agencies merged into Office of War Information, 44; American, 1945–1946, 12–13; American, 1947–1949, 16–18; Psychological Warfare Division creates apparatus for, 9
Information Control Division (ICD): becomes Information Services Division, 121; Documentary Film Unit, 18, 57–60; establishment of, 9, 12; in film policy, 41, 48, 49, 50; fine arts as blind spot of, 69–87; fine arts policy of, 74–76; personnel changes in, 17; press policy of, 108–9, 126, 167n18; on *Todesmühlen*, 55, 57; two aspects of policy of, 12–13; and *Ulenspiegel*, 111, 116–17, 119, 120; Vigorous Information Program of, 119
Information Control Review, 39, 58, 101

information films, 46
Information Services Division (ISD), 121
"In Grunewald Forest" (Sandberg), 123–24
Insulaner, Der (cabaret), 121
Insulaner, Der: Das Magazin für das Insel-Leben (magazine), 121, 129
intelligence, American. *See* American intelligence
International Military Tribunal (Nuremberg), 4, 36, 37, 92
"invasion films," 48
"iron curtain" speech, 115
Ivens, Joris, 45

Jackson, Robert H., 37
Jaenisch, Hans, 161n7
Japanese Relocation (film), 46
Jews. *See* anti-Semitism; Holocaust
Joint Chiefs of Staff Directive 1067 (JCS 1067), 11–12, 13, 17, 22, 74, 106–7
Joint Chiefs of Staff Directive 1779 (JCS 1779), 17, 101
Joint Communiqué (Yalta Conference), 106
Joseph, Robert, 49
Josselson, Michael, 56, 133
Jud Süss (film), 47

Kahnweiler, Daniel-Henry, 71
Kallweit, Ismar, 113
Kammer der Kunstschaffenden, 15, 76
Kampfbund für Deutsche Kultur, 72
Katyn massacre, 34
Kaus, Max, 161n7
Kennan, George, 116
Keyes, Geoffrey, 61, 62
Kilger, Heinrich, 110, 113
Kipling, Rudyard, 42
Kirchner, Ernst Ludwig, 77
Kirstein, Lincoln, 76, 159n43, 161n13
Klee, Paul, 77
Kline, Herbert, 52

Know Your Job in Germany (film), 26
Koch, Mimi Killian, 98
Koerner, Henry, 92–95, 161n13
Koestler, Arthur, 19
Kokoschka, Oskar, 77
Kolbe, Georg, 161n7
Kollwitz, Käthe, 77, 112, 161n13
Konnerth, Lotte, 98
Kracauer, Siegfried, 47
Kreische, Gerhard, 56, 114
Kukryniksy, 110
Kultur, 13, 101–3, 116, 130, 164n46
Kulturbund zur demokratischen Enneuerung Deutschlands, 15, 76, 168n23
Kunst (magazine), 74
Kunst dem Volk (magazine), 74
Kunst im Deutschen Reich, Die (magazine), 74
KZ: A Pictorial Report from Five Concentration Camps (Office of War Information), 33–34, 35, 51

Land, The (film), 43
"Land-Land! Deutsch-land!" (Shaw), 124
Lange, Dorothea, 43
Langer, William, 7
Lardner, Ring, Jr., 64, 67
Lasky, Melvin J., 17, 121, 133
Lautlose Aufstand, Der: Bericht über die Widerstandsbewegung des deutschen Volkes, 1933–1945 (Weisenborn), 120
Law No. 52, 107
Lebenspiegel (Koerner), 93
Lehmann-Haupt, Hellmut: on antimodernism, 81–85, 87, 159n43; as Art Intelligence Unit head, 94; background of, 82; German-American cultural consolidation recommended by, 85–86; in incorporation of fine arts into American cultural propaganda, 87; and Koerner exhibit, 92, 161n13; in Prolog group, 90

Lehmbruck, Wilhelm, 77
Lest We Forget (exhibit), 25–27, 28
Leube, Max, 161n7
Ley, Robert, 72
Leymarie, Jean, 99
Life (magazine), 24, 26, 27, 44, 93, 169n34
Linebarger, Paul, 50–51
Linfert, Karl, 113
Linton, Leonard, 22, 141n9
literature, 12, 15, 19
Litvak, Anatole, 45
Lommer, Horst, 111
"Long-Range Policy Statement for German Re-education" (OMGUS), 101
Look (magazine), 27
Lorentz, Pare, 43, 58, 62, 145n5
Los Angeles Art Club, 79
Los Angeles Examiner (newspaper), 80
Lubitsch, Ernst, 45
Luce, Henry R., 43–44
Luckner, Graf, 161n7
Ludwig of Bavaria, King, 70
Lumière, Auguste and Louis, 42
Lusset, Félix, 20
Lutzeier, Paul, 90, 160n2

Magazine of Art, 97
Mahoney, Haynes R., 50
Maltese Falcon, The (film), 49
March of Time, The (film series), 44
Marcks, Gerhard, 161n7
Marin, John, 78
Marshall, George C., 17, 24, 45, 80, 116
Marshall Plan, 17, 58, 60, 116, 122, 123, 124, 151n71
Marshall Plan Economic Cooperation Administration Film Unit, 58
Martin, David Stone, 92
Martin, Kurt, 99
Masereel, Frans, 110
Mataré, Ewald, 99
May, Andrew J., 63
McClaskey, Beryl, 90, 91, 96
McCloy, John J., 8
McClure, Robert A., 8, 9, 12, 49, 52
McKelway, Benjamin M., 26, 142n18
McKisin, John, 24
McKnight, Eline, 96
McMahon, B. B., 66
media: American conservatives employ, 81; become anticommunist propaganda tool, 17. *See also* film; press, the
Meistermann, George, 100
Melzer, Karl, 161n13
Merbitz, Bruno, 161n7
MFA&A. *See* Monuments, Fine Arts & Archives Section (MFA&A)
Military Government Regulation #1, 87
Military Government Regulation Title 18-401.5, 107
Military Intelligence Division (MID), 6–7, 8
Miller, Irving, 92
Miller, Lee, 25
Ministerium für Volksaufklärung und Propaganda, 46
Mnookin, Jennifer, 27
modern art: American conservative opposition to, 70, 78–81, 97, 103; American support for, 5, 18, 89; *Ausstellung: Henry Koerner U.S.A. 1945–1947* exhibit, 92–95, 161n13; Blevins Davis Prize, 99–100; flourishes in Germany, 71; "Formalism and Realism in the Fine Arts" debate, 125; French military government support of, 20; German conservative opposition to, 70; Lehmann-Haupt on antimodernism, 81–85; Nazi purging of, 3, 4, 69, 71–73; Prolog group and, 90; resurgence in Germany after 1945, 89, 102–3, 128; Sedlmayr's opposition to, 100; Soviet denigration of, 5, 16, 18, 77, 81, 83–84, 103, 118, 123; in *Ulenspiegel*, 110; *Zeitgenössische Kunst and Kunstpflege in U.S.A.* exhibit, 97–99

202 Index

Moholy-Nagy, Lázló, 163n33
Moll, Margarethe, 161n7
Monat, Der (journal), 121
Montgomery, Bernard Law, 48
Monthly Information Bulletin (HICOG), 50
Monthly Report of Military Governor U.S. Zone, 57
Monuments, Fine Arts & Archives Section (MFA&A), 75–76; Art Intelligence Unit of, 94; *Ausstellung: Henry Koerner U.S.A. 1945–1947* exhibit, 92–95; background of personnel of, 75; in cultural preservation, 75, 108; fine arts policy developed by, 69–70, 81–87; in iconoclasm campaign, 106; postwar role of, 75
Morgenthau, Henry, Jr., 11, 22, 59
Motion Picture Producers and Distributors of America, 43, 105
motion pictures. *See* film
Müller, Rosemarie, 161n13
Munsing, Stefan, 99, 100
Münzenberg, Willi, 139n30
museums, 31, 70, 71, 74, 75

National Board of Review, 105
National Day of German Art, 74
National Gallery (Berlin), 71
nativism, 6
Nazi Concentration Camps, The (film), 36
Nazis: American propaganda shifts away from, 18; art as propaganda for, 69, 83; art stolen by, 69, 75, 87, 107; Benedict and Weltfish's *The Races of Mankind* refutes racial theory of, 62; changing American narratives of, 2; Communist interpretation of Nazism, 14; Documentary Film Unit anti-Nazi films, 58–59; in East Germany, 123; film used as propaganda, 46–47; Goebbels, 34, 46, 72, 74; Göring, 59, 72; Military Intelligence Division attitude toward, 6; modern art purged by, 3, 4, 69, 71–73; OMGUS iconoclasm campaign against visual legacy of, 4, 106–8, 166n7; positive fine arts policy of, 73–74; reenter West German political and economic life, 18; *Ulenspiegel* on, 113; visual propaganda of, 3; *Weltanschauungskrieg* concept of, 8; West Germans continue to have high opinion of, 131; younger Germans initiate collective discussion about, 132. *See also* concentration camps; denazification; Hitler, Adolf
Nazis Strike, The (film), 45
Negro Newspaper Publishers Association, 61
Nerlinger, Oscar, 110, 113, 114
Netzband, Georg, 162n23
Neue Adam, Der (Meistermann), 100
Neue Sachlichkeit (New Objectivity), 73, 93, 155n6
Neues Deutschland (periodical), 116, 125
Neue Zeitung (newspaper), 115–16, 117, 119
Neumann, Günter, 121, 129
New Deal: documentary films about, 41, 43; Information Control Division personnel influenced by, 108; photography, 27; Republicans begin fight to undo, 5
New German Art, 73–74, 157n26
New Objectivity (Neue Sachlichkeit), 73, 93, 155n6
New School for Social Research, 8
newsreels, 42, 44, 45, 48, 50
Newsweek (magazine), 25
"New Vision" movement, 163n33
New York Journal (newspaper), 79
Nicht stören! Funktionärsversammlung (film), 121
Noce, Daniel, 64–65, 154n91
nonfiction films: functions of, 42; news-

reels, 42, 44, 45, 48, 50; in World War II, 44–46. See also documentary films
nonfraternization order, 22, 23, 36–37
Nordau, Max, 70
Nuremberg trials (International Military Tribunal), 4, 36, 37, 92

obedience training, 9
Occupation (handbook), 37–38
occupied Germany: American atrocity propaganda in, 21–39; American control of information, 1945–1946, 12–13; American control of information, 1947–1949, 16–18; American film policy in, 46–51; American occupation of Germany, 9–12; atrocity propaganda's effect in, 28–39; as Cold War battleground, 1; constants in American policy in, 127; cultural initiatives in British zone, 18–19; cultural initiatives in French zone, 19–20; first free elections in West Germany, 130; initial impact of atrocity propaganda in, 32–35; liaisons of German women and American soldiers, 37; politics and culture in Soviet zone, 13–16; quadripartite rule in, 1, 10; United States affected by occupation, 132–33; United States and Britain consolidate their zones, 115; West Germany not Americanized, 131
Office of Censorship, 44, 105
Office of International Information and Cultural Affairs (OIC), 78
Office of Military Government U.S. See OMGUS (Office of Military Government U.S.)
Office of Strategic Services (OSS), 6–8
Office of War Information (OWI): censorship role of, 105; functions of, 3; *Government Information Manual for the Motion Picture Industry*, 44–45; Hollywood's relations with, 3; information-control agencies merged into, 44; *KZ: A Pictorial Report from Five Concentration Camps*, 33; nonfiction films produced by, 46, 146n13; personnel in Psychological Warfare Division, 8
O'Keeffe, Georgia, 78
OMGUS (Office of Military Government U.S.): anticommunist campaign of, 4, 17, 116, 128, 130–31; changing American narratives of German fascism, 2; Cold War affects denazification policy of, 122; Cold War affects press policy of, 115–17, 128–29; Cold War gives German artists bargaining power with, 95–97; cultural competition with SMAD, 95; diversity of, 13, 41; dual mission of, 10; establishment of, 10; film used by, 4, 13, 41, 47–51, 128; *Fragebogen* (political questionnaire) used by, 109, 113; German film attendance monitored by, 56, 150n53; iconoclasm campaign of, 4, 105, 106–8, 128; Information Control Division incorporated in, 12; Joint Chiefs of Staff Directive 1067 informs policy of, 11–12; *Kultur* promoted by, 101–3; media used to disseminate message of, 1; on *Der Monat,* 121; *Neue Zeitung* launched by, 115; new anticommunist agenda of, 17–18; as nondemocratic government, 127; political factors influence, 5, 81; racial tensions within, 61; radicalizes cultural policy, 116; tightens control of German press, 121; *Todesmühlen* documentary, 21–22, 51–57; victory in psychological warfare of, 130–31; *Zeitgenössische Kunst and Kunstpflege in U.S.A.* exhibit authorized by, 97–99. See also Information Control Division (ICD); Monuments, Fine Arts & Archives Section (MFA&A)

Operation Eclipse, 137n16
Operation Talk Back, 17, 117
Orlowski, Hans, 161n7
Orwell, George, 19
OSS (Office of Strategic Services), 6–8
OWI. *See* Office of War Information (OWI)

Patterson, Robert P., 154n91
Patterson, William D., 148n41
Patton, George, 21, 22, 30
Pechstein, Max, 77, 113
Perlin, Bernard, 92
photography: concentration camp photographs, 21, 24, 25–26, 39, 51; falsification of, 142n23; in Goebbels's *Amtliches Material zum Massenmord von Katyn*, 34; in *KZ: A Pictorial Report from Five Concentration Camps*, 33, 34; "New Vision" movement, 163n33; at Nuremberg trials, 36; objectivity attributed to, 26; political, 27; war, 27
Picasso, Pablo, 20, 112
Picture Post (magazine), 27
Pilgert, Henry P., 49–50
Plow That Broke the Plains, The (film), 43
Political Intelligence Division of the Foreign Office, 19
"Political Project: Graphic Artists in Berlin," 96
Poor, Henry Varnum, 99
Porgy and Bess (Gershwin), 164n37
Power and the Land (film), 43
Prelude to War (film), 45
President's Commission on Civil Rights, 65
press, the: Americans encourage democratic German, 13, 108–9, 126, 167n18; censorship of, 12, 108; Cold War and change of American policy regarding, 115–17, 126, 128–29; in OMGUS anticommunist campaign, 17–18; OMGUS tightens control over, 121; *Ulenspiegel* case, 110–26
Price, Byron, 36, 144n45
Production Code Administration, 43, 44, 105
Prolog, 89–92, 96, 161n7
propaganda: American anticommunist, 17–18; American films as, 4, 13; American propaganda films, 41–67; art as, 69, 83, 97; "black," 7, 9; documentary films as, 42–44; Goebbels's *Amtliches Material zum Massenmord von Katyn*, 34; more forward-looking message for American, 36; Nazi propaganda films, 46–47; photography in, 27; racism as issue in anti-American, 62, 65–66, 129, 151n71; Soviet anti-American and anticapitalist, 16, 18; Soviet approach to German guilt, 34–35; Wehrmacht propaganda companies, 47. *See also* atrocity propaganda; strategic propaganda; visual propaganda
Prophet, Der (Koerner), 93
Psychological Warfare Division (PWD), 8–9; *Atrocities: A Study of German Reactions*, 33, 34, 145n6; civilian contingent of, 6–7; photographs of concentration camps ordered by, 51; role in occupied Germany, 9; *Todesmühlen* produced by, 52
Psychological Warfare School, 9
public burials, 31–32
Pulitzer, Joseph, 25–26, 30, 142n18
PWD. *See* Psychological Warfare Division (PWD)

race: *The Brotherhood of Man* on, 61–67. *See also* racism
Races of Mankind, The (Benedict and Weltfish), 62–63, 65, 67, 152n75
racism: of Bilbo, 152n78; documentary films for export ignore, 46; as issue

in anti-American propaganda, 62, 65–66, 129, 151n71; of Military Intelligence Division, 6; as threat to American credibility, 61, 65. *See also* anti-Semitism
Radio Berlin, 35
Radio in the American Sector (RIAS), 119, 121
Radler, Max, 162n23
Rajk, Lazlo, 124, 171n55
Rapf, Maurice, 64
Rebay, Hilla von, 98–99
Redslob, Edwin, 94, 113
Regionalism, American, 157n26
Reichskulturkammer, 72
Report to the American People, A (Pulitzer), 30
Resettlement Administration (RA), 43
Reuther, Walter P., 62, 64–66
Revolution der modernen Kunst, Die (Sedlmayr), 100
Riefenstahl, Leni, 56
Right to Strike, The (film), 49
"Right Way, The" (Holtz), 119
Rilla, Paul, 111
Ritter, John, 96
River, The (film), 43
Robb, Marilyn, 79
Roberts, Owen J., 75
Roberts Commission (American Commission for the Protection and Salvage of Artistic and Historic Monuments), 75
Roh, Franz, 98, 163n33
Rorimer, James, 76
Rosen, Gerd, 98
Rosenberg, Alfred, 72
Rosenberg, Sam, 92, 161n13
Rosie, Paul, 161n7
Ross, Marvin C., 75
Rothschilds, Die (film), 47
Royall, Kenneth Claiborne, 65, 66, 154n91

Ruf, Der (journal), 117
Ruhtenberg, Cornelis, 161n7
Rust, Bernhard, 72

Saltonstall, Leverett, 26, 141n17
Sandberg, Herbert, 110; art for *Ulenspiegel* by, 113–14, 118–19, 123–24, 162n23; takes *Ulenspiegel* to East Berlin, 121, 124–25; as *Ulenspiegel* art editor, 110–11, 112, 126
Sartre, Jean-Paul, 20, 112, 113, 121
satirical journals, 167n19
Schäfer-Ast, Albert, 110
Schirach, Baldur von, 72
Schlichter, Rudolf, 110, 113, 162n23
Schmidt-Rottluff, Karl, 77, 161n7
Schroder-Sonnenstern, Friedrich, 162n23
Schulberg, Budd, 57
Schulberg, Stuart, 57–58
Schulenberg, Tisa von, 162n23
Schultze-Naumburg, Paul, 72
Schumacher, Kurt, 130
Schutzverband deutscher Schriftsteller, 111
Schwitters, Kurt, 163n33
SED (Sozialistische Einheitspartei Deutschlands; Socialist Unity Party), 15, 59, 76, 77, 112, 118, 123, 125
Sedlmayr, Hans, 100
Seiller, Edward F., 30–31, 32
Seitz, Gustav, 161n7
SHAEF (Supreme Headquarters, Allied Expeditionary Force), 137n17
Shahn, Ben, 43, 78, 92, 93, 158n39
Shaw, Elizabeth, 124
Short, Dewey, 26, 142n17
Sieg des Glaubens, Der (film), 56
Signal (magazine), 27
Sikes, Robert L. F., 63
Simplicissimus (journal), 110
Sintenis, Renée, 96, 161n7
Skutsch, Karl Ludwig, 161n13

SMAD (Sowjetische Militäradministration in Deutschland; Soviet Military Administration, Germany): anti–modern art campaign of, 77, 81, 83–84, 103, 118, 123; cultural competition with OMGUS, 95; cultural warfare policy of, 4–5, 14–16; film used by, 4; former Nazis in, 123; *Frischen Wind* as satirical journal of, 123; mass organizations created by, 15; media used to disseminate message of, 1; Prolog group as counter to, 90; radicalizes cultural policy, 116; in revival of German fine arts, 76–77; *Der Ruf* denounced by, 117; Sartre attacked by, 111; *Tägliche Rundschau*, 15, 16; as targets of American propaganda, 17–18; *Ulenspiegel* paper supply cut by, 125

Smyth, Craig Hugh, 76, 86–87

Social Democrats, 132

Socialist Realism, 5, 77, 83, 125

Socialist Unity Party (Sozialistische Einheitspartei Deutschlands; SED), 15, 59, 76, 77, 112, 118, 123, 125

Sontag, Susan, 26

"Soviet Art and Its Relation to Bourgeois Art" (Dymschitz), 81

Soviet Union: American permissiveness exploited by, 126; anti-Semitic offensive of, 16, 123, 124; arts policy in occupied Germany, 15, 16, 69, 76, 97; Berlin sector of, 10; consolidates hold over Eastern Europe, 16, 116; denazification under, 123; film employed in occupied Germany by, 15, 47–48; fine arts in Soviet zone and sector, 76–77; Free Trade Union (FDGB) controlled by, 95, 96; on German guilt, 34–35; Marshall Plan denounced by, 17, 116, 122; modality of German nationalism of, 119; modern art opposed by, 5, 16, 18, 77, 81, 83–84, 103, 118, 123; politics and culture in Soviet-occupied Germany, 4–5, 13–16; race as issue in Cold War propaganda of, 62; Stalin, 16, 34, 77, 106, 116, 126; *Ulenspiegel* becomes platform for Soviet propaganda, 120–21; *Ulenspiegel* on, 118, 122; zone of occupation of, 10. *See also* Cold War; SMAD (Sowjetische Militäradministration in Deutschland; Soviet Military Administration, Germany)

Sozialistische Einheitspartei Deutschlands (SED; Socialist Unity Party), 15, 59, 76, 77, 112, 118, 123, 125

Spanish Civil War, 27, 43, 108

Speer, Albert, 72

Spiegel, Der (magazine), 19

Stabenau, Friedrich, 161n7

Stalin, Joseph, 16, 34, 77, 106, 116, 126

Standen, Edith Appleton, 76, 82, 86–87

Stangl, Otto, 98

Stars and Stripes (newspaper), 22, 26

Staudinger, Karl, 162n23

Stewart, Jack M., 169n34

Stimson, Henry, 11

strategic propaganda: as Cold War factor, 1; cultural policy as, 1–2; focuses on anticommunism, 17; not antagonizing its target, 36, 56

Strecker, Paul, 161n7

Strempel, Horst, 162n23

Strong, George V., 7

student revolt of 1968, 132

Sturm, Der (journal), 71

"Supreme Headquarters Official List of Protected Monuments," 75

Sweet, Frederick A., 86

Szluk, Peter F., 91–92

Taber, John, 80

Tagesspiegel, Der (newspaper), 94

Tägliche Rundschau (newspaper), 15, 16, 77, 81, 116, 123

Taylor, Davidson, 52, 148n41
Tendenzfilme, 46–47
Tennessee Valley Authority (film), 49
Textor, Gordon E., 116, 117
theater, 12, 15
Third Way, 114–15, 120
"This is Democracy" (Sandberg), 124
"This Is Why We Fight" (pamphlet), 23
Thompson, Robert H., 25
"Three Holy Kings, The" (Sandberg), 113
Time (magazine), 25, 44, 93, 169n34
Tiomkin, Dimitri, 45
Tiul'panov, Sergei, 14–15, 16, 77
Todesmühlen (documentary film), 21–22, 51–57
Town, The (film), 49
training films, 3, 45
Truman, Harry S.: and *Advancing American Art* exhibit, 80; and *Brotherhood of Man* film, 64–65, 66; conservatives constrain policy of, 129; declares ideological war against communism, 16–17; desegregation of the military by, 66, 154n91; Executive Order 9835, 16, 116; Executive Order 9981, 154n91; film of concentration camps seen by, 24; Soviet Union denounced by, 126

Ufa (Universum-Film-Aktiengesellschaft), 46–47
Uhde, Wilhelm, 71
Ukrainian Daily News, 64
Ulbricht, Walter, 14, 112, 125, 169n26
Ulenspiegel (magazine), 110–26; on "bad Germans" and "good Germans," 114; becomes platform for Soviet propaganda, 120–21, 123–25; bias of, 117–19, 121–23; circulation cut by Information Control Division, 120, 129; circulation of, 112; closing of, 125; Cold War and American support for, 119–23, 128–29; Communist press compared with, 122; in East Berlin, 121, 123–25; headquarters as meeting place for anti-Nazi intellectuals, 112–13; meaning of name, 167n19; Michel as figure of, 113–14; in 1945 and 1946, 112–15; in 1948, 119–23; as nonconventional Soviet asset, 126; progressive nationalism advocated by, 119–20; on Soviet Union, 118, 122; Third Way perspective of, 114–15; on *Todesmühlen,* 56
United Productions of America (UPA), 153n81
United States: affected by occupation of Germany, 132–33; American occupation of Germany, 9–12; atrocity propaganda's effect on American home front, 22–28; Berlin sector of, 10; and Britain consolidate their zones, 115; conservative opposition to modern art in, 70, 78–81; desegregation of the military, 61–62, 66, 154n91; domestic disagreement over German occupation policy, 11; Marshall Plan, 17, 58, 60, 116, 122, 123, 124, 151n71; modality of German nationalism of, 119; racism as issue in anti-American propaganda, 62, 65–66, 129, 151n71; zone of occupation of, 10. *See also* American intelligence; Cold War; Office of War Information (OWI); OMGUS (Office of Military Government U.S.); Truman, Harry S.
United States Film Service (USFS), 43
United States Information Centers (Amerikahäuser), 99, 102, 119, 163n34
U.S. Feature Service (Amerika Dienst), 121
U.S. Forces, European Theater (USFET), 137n17
U.S. Office of Censorship, 44, 105

USO (United Services Organizations), 63, 64

Variety (newspaper), 47
Veiller, Anthony, 45
Vercors (Jean-Marcel Bruller), 20
Verists, 71, 73, 155n6
Verlust der Mitte (Sedlmayr), 100
"Very Small and the Voracious Moloch, The" (Kilger), 113
Vetter, Ewald, 161n13
Vigorous Information Program (VIP), 119
visual propaganda: American art policy, 4–5; atrocity propaganda as, 3–4, 128, 129; changing American narratives of German fascism, 2; about concentration camps, 3–4, 21–39; in denazification and reeducation, 128; factors shaping American, 5; importance of, 2–3; Nazi, 3; of Office of War Information, 3; for selling private enterprise and democracy, 4. *See also* censorship; fine arts; film; iconoclasm; photography; press, the
Vogue (magazine), 25

Wallenberg, Hans, 116, 169n34
War Comes to America (film), 45
War Production Board, 45
Washam, Ben, 64
Weekly Information Bulletin (OMGUS), 57
Weg, Der (Sandberg), 125
"We Have Always Been Anti-fascists" (Sandberg), 113
Weidler, Charlotte, 97
Weimar Republic, 71

Weir, John M., 25
Weisenborn, Günther, 110, 111–12, 120, 126
Weltanschauungskrieg, 8
Weltfish, Gene, 62–63, 63–63, 67, 151n71
Welt im Film (newsreel), 21, 55
Werner, Theodor, 99
Westintegration, 122
Why We Fight series, 45
"*Wiedergutmachung*" (Sandberg), 118
Wilder, Billy, 53–54, 55
Wilhelm II, Kaiser, 70
Winkler, Friedrich, 90–91, 161n7
Winter, Fritz, 99
"With Other Eyes" (Effel), 118
Wölfflin, Heinrich, 70
World at War (film), 46
World War I: censorship, 105; photography, 27
Worringer, Wilhelm W., 70, 72
Wuellfarth, Leonhard, 100

Yalta Conference Joint Communiqué, 106
Yank (weekly), 23

Zeitgenössische Kunst and Kunstpflege in U.S.A. (exhibit), 97–99
Zeller, Magnus, 162n23
ZEN 49, 99
Zhdanov, Andrei, 16, 77, 123
Ziegler, Adolf, 72–73
Zimmermann, Mac, 161n7
Zink, Harold, 1–2, 41
Zion, Ben, 78
Zwei Städte (film), 60
Zwischen West und Ost (film), 60